Matrix and digital computer methods in structural analysis

Consulting Editor
Professor P. B. Morice
University of Southampton

Other titles in the European Civil Engineering Series

Johnson: Structural Concrete
Pilcher: Principles of Construction Management
Schofield and Wroth: Critical State Soil Mechanics
Zienkiewicz: The Finite Element Method in Structural and Continuum Mechanics

Matrix and digital computer methods in structural analysis

W. M. Jenkins

Head of Department of Civil and Structural Engineering and Building,
Constantine College of Technology, Middlesbrough.
Formerly Reader in Civil Engineering in the University of Bradford.

McGRAW-HILL · LONDON

New York · St. Louis · San Francisco · Sydney · Toronto · Mexico
Johannesburg · Panama

Published by
McGRAW-HILL Publishing Company Limited
MAIDENHEAD · BERKSHIRE · ENGLAND

07 094112 2

PRINTED AND BOUND IN GREAT BRITAIN

Preface

The developments in methods of structural analysis which have taken place during the last decade have been due almost entirely to the impact of the digital computer. Although the principles of equilibrium and compatibility, and the relationships between forces and displacements, remain basic in all the methods, their interpretation, within the various numerical processes, has received much attention. Freedom from the necessity to economize on arithmetic in order to obtain a reasonably quick solution, has enabled the structural analyst to devote his attention to more accurate methods of handling large complex structures. An outstanding example of this development is the 'finite element' technique which has gained popularity at a prodigious rate in recent years. The digital computer is a prerequisite in the application of the finite element method.

A highly organized method of calculation is essential when using a digital computer; thus, matrix formulations of analytical methods have an important place in the overall development. The advantages of using matrix methods with the digital computer are two-fold: first, the storage of data, in matrix form as arrays of numbers, is natural to the computer; second, the operations of matrix algebra can be provided in the form of standard routines which are readily available and can be brought into operation by compact instructions in the program.

The digital computer and matrix methods should be regarded as tools which are available for use when required. Although matrix methods are very suitable for digital computer work, they are not usually the best choice for hand calculations. However, some experience of hand calculations, using matrix methods, is advisable in order to gain familiarity with the methods before going on to the computer. The reader is expected to have a working knowledge of elementary matrix algebra. Most undergraduate courses in engineering mathematics contain an adequate treatment of matrices. The operations of addition, subtraction, multiplication, transposition, and inversion, together with the ability to manipulate matrices in partitioned form, are required by the reader.

This book is intended to provide a reasonably comprehensive treatment of matrix and digital computer methods in structural analysis, at a level suitable for final year students in structures and students in postgraduate courses in structures. In most undergraduate courses the classical methods of analysis and structural theorems are generally well covered by the end of the second of a three-year course, and a substantial part of the final year is directed at the more recent developments in matrix methods and the use of digital computers. It is hoped that this book will meet the requirements of the undergraduate in this respect, yet contain enough advanced material, with examples of practical applications, to meet the needs of the postgraduate student and practising engineer. The treatment is not intended to be exhaustive or particularly advanced, but for the reader wishing to study particular aspects in greater depth, it should serve as an introduction to more advanced texts, such as those by Zienkiewicz[42] and Przemieniecki[31].

The matrix methods of structural analysis divide broadly into flexibility and stiffness methods. The basic concepts relevant to both methods are discussed in chapter 1. Chapters 2 and 3 are devoted to the flexibility method and 4 and 5 to the stiffness method. In both cases, the basic method is outlined in the first chapter of the pair and further applications and refinements considered in the second. Chapter 6 is devoted to the finite element method, and is intended to serve as an introduction to this important and rapidly developing method.

Chapter 7 deals briefly with digital computers and programming and then with the important topic of numerical methods for band matrix solution. The band matrix occupies an important position in the field of structural analysis by digital computer, and it is desirable to have efficient computer programs for the solution of a general symmetric band matrix. For this purpose, two complete programs are given at the end of this chapter, one in ALGOL and the other in FORTRAN.

Chapter 8 consists of a selection of examples of computer applications, and here the intention is to suggest approaches to the organization of the methods for particular types of structure.

Some problems are included in the text to enable the reader to gain experience with the methods. It is intended that matrix methods will be used to obtain solutions, but it is to be understood that such methods are not necessarily the best choice for computations by hand. Answers to the problems are provided at the end of the book. The author wishes to express his gratitude to the University of Bradford and to the Institution of Civil Engineers for permission to use questions from examination papers set by these bodies.

Much of the material for this book originated in lecture courses given by the author to final year classes and to postgraduate short courses at the University of London, King's College, and at the University of Bradford. The author acknowledges his gratitude to present and former colleagues and students who have helped by reading the manuscript and checking the examples. In particular, the author would like to thank J. M. Siddall, A. G. Bellamy, D. J. Lax, and D. H. Little. The development of the anisotropic, parallelogrammic flexural element in chapter 6 was carried out by J. M. Siddall. The FORTRAN band matrix solution program by Gauss elimination is based on the original EMA program written by the author's former research assistant, Mrs. J. M. Towers.

A special debt of gratitude is due to the author's research assistant G. C. de Jesus, who contributed in many ways to the preparation of the manuscript and made several valuable suggestions and contributions, including the band matrix solution program using the Choleski method.

Finally, the author appreciates the help given by Miss Sylvia Wilman in undertaking the difficult task of typing the manuscript.

W. M. JENKINS

Units

In the numerical examples and problems, the new SI metric units have been adopted (B.S. 3763:1964 The International System (SI) units). In this system, the basic units of length and mass are the metre (m) and kilogramme (kg) respectively. The unit of force is the newton (N) and is that force which, when applied to a body having a mass of 1 kilogramme, gives it an acceleration of one metre per second squared. For structural work, the newton is inconveniently small (0·225 lbf), and the kilonewton (kN) (1000 N, 225 lbf) will be used generally.

When dealing with gravity loading on structures it will be necessary to convert the unit weights of material from kgf/m^3 units to N/m^3, the required relationship being

$$1 \text{ kgf} = 9{\cdot}81 \text{ N} \quad (9{\cdot}80665 \text{ exactly})$$

All other applied forces are quoted directly in N or kN.

Other units adopted are

Bending moment	kNm
Stress	N/mm^2
Young's modulus (30 × 10^6 lbf/in^2 = 207 kN/mm^2)	kN/mm^2
Second moment of area	mm^4

The mm is used with the metre, in preference to the cm, since it is preferable to work consistently with powers of ten which are multiples of ± 3.

The following conversion factors may be useful to the reader who is unfamiliar with SI units:

Non-metric	SI	Reciprocal
1 in	25·4 mm	0·0394
1 ft	0·3048 m	3·2808
1 lbf	4·448 N	0·2248
1 lbf/in^2	$6{\cdot}895 \times 10^{-3}$ N/mm^2	145·03
1 lbf/ft^2	$47{\cdot}88 \times 10^{-3}$ kN/m^2	20·89
1 tonf/in^2	15·44 N/mm^2	0·0648
1 tonf/ft^2	107·3 kN/m^2	$9{\cdot}32 \times 10^{-3}$
1 lbf in	113×10^{-6} kNm	$8{\cdot}85 \times 10^{3}$
1 tonf in	253×10^{-3} kNm	3·95
1 lbf ft	$1{\cdot}36 \times 10^{-3}$ kNm	736
1 tonf ft	3·04 kNm	0·329

Notation

A, **B**	Displacement transformation matrix
D	Elasticity matrix
f	Element flexibility matrix
F	Structure flexibility matrix
k	Element stiffness matrix
K	Structure stiffness matrix
m_0	Statically determinate bending moment distribution due to applied loads
M	Bending moment
M_x	Twisting moment
$\mathbf{M}^T$	Transposed matrix
N	Axial force
Q	Stress matrix
r, **s**	Nodal displacement matrix
R	Applied load matrix
S, **Y**	Matrix of element nodal forces
T	Transformation matrix
U	Strain energy
u, **U**	Primary structure displacements
V, v_i	Shear force
x, **X**	Redundant forces
θ	Rotation
ν	Poisson's ratio
u, v, w	Linear displacements
[ ]	Rectangular matrix
{ }	Column matrix (written in 'row' form)
⌈ ⌋	Diagonal matrix

Contents

1. Basic Concepts

An essential part of the process of structural design is the assessment of the response of a given structure to the application of forces and other disturbances, such as change in temperature and movement of supports. The structural response is normally described in terms of changes in the state of stress or strain of the structure. The internal distribution of stress gives rise to 'stress resultant' actions such as bending moment, shear force, thrust, and torque, and these are collectively termed 'forces'. In considering the deformations of the structure, attention may be focused on the linear or angular displacements occurring at discrete points in the structure.

In framed structures, it is helpful and convenient to describe the state of stress and deformation in terms of forces and displacements at selected points, termed 'nodes', in the structure. These are usually the ends of members or the joints. This process introduces the concept of a structural element; a component part of a structure, such as a strut or a beam. The behaviour of the structural element, in the form of force-displacement relationships, is usually already known or can be obtained without difficulty. In chapter 6, more sophisticated elements will be considered and although the determination of the force-displacement relationships of such elements is a major exercise it has to be done only once for a typical element.

The structural analyst uses such elements in building a complete picture of the behaviour of the structure. He uses the element properties to assemble a mathematical model of the structure and describe its properties in the assembled condition. Once this has been done, the response of the complete structure to a set of applied loads or other disturbances can be determined. In the process of assembling the structure from its individual components, the conditions of geometrical compatibility must be satisfied. The deformations of the elements under load must be such that, on assembly, they will fit together. In addition to this, the conditions of statical equilibrium must be satisfied; each element must be in equilibrium under the action of all the forces acting on it, both external loads and reactions, and forces coming from adjacent elements.

It is, thus, the purpose of structural analysis to determine the state of deformation or stress of the structure, based on known force-displacement characteristics of the component parts, and satisfying all the conditions of compatibility and equilibrium.

1.1. Force-displacement Relationships

These relationships are known for standard structural elements, and in other cases can be derived by established procedures depending on the known geometry of the element and the elastic properties of the material. The relationships enable forces to be calculated from a knowledge of displacements, or displacements from a knowledge of the forces.

Two simple structural elements are shown in Fig. 1.1, along with their force vectors and corresponding displacement vectors. If the force vector is

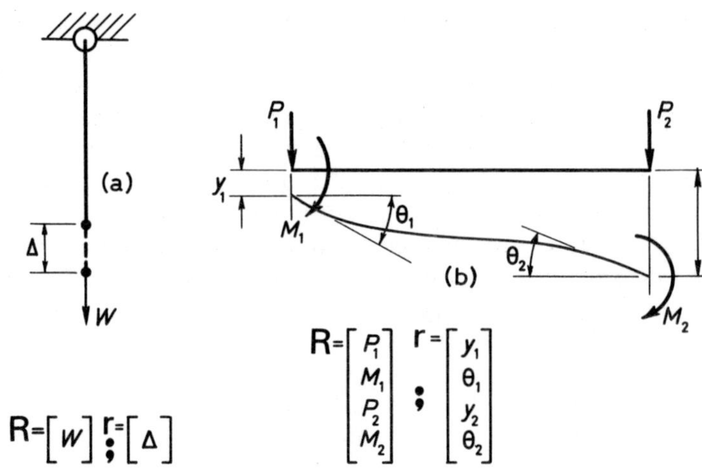

Figure 1.1. Force and corresponding displacement vectors

denoted by **R** and the displacement vector by **r**, then, in general, there will be a relationship

$$\mathbf{R} = \mathbf{k}\mathbf{r} \tag{1.1}$$

where the matrix **k** contains the coefficients of the displacements in the force-displacement relationships. These relationships may be expressed in the inverse form:

$$\mathbf{r} = \mathbf{f}\mathbf{R} \tag{1.2}$$

in which the matrix **f** contains the coefficients of the nodal forces in the displacement-force relationships. Thus **k** and **f** are related,

$$\mathbf{k} = \mathbf{f}^{-1} \tag{1.3}$$

The force vector **R** may contain a number of different types of forces, bending moments, twisting moments, shear forces, and thrusts, and the displacement vector **r** will contain the corresponding displacements in corresponding positions. For example, if the force in the ith position of the force vector is a moment M_j at position j in the structural element, then the displacement in the ith position of the displacement vector will be θ_j, the rotation at j in the same direction as the moment M_j.

Linearity. If the coefficients in the matrices **k** and **f** of eqns. (1.1) and (1.2) are constants, independent of the applied loads, then the force-displacement relationships are linear. It follows that the total effect of a number of loads can be obtained by simple addition of the effects of the separate loads. This is the *principle of superposition*. The majority of structural problems involve a linear relationship between force and displacement, and the principle of

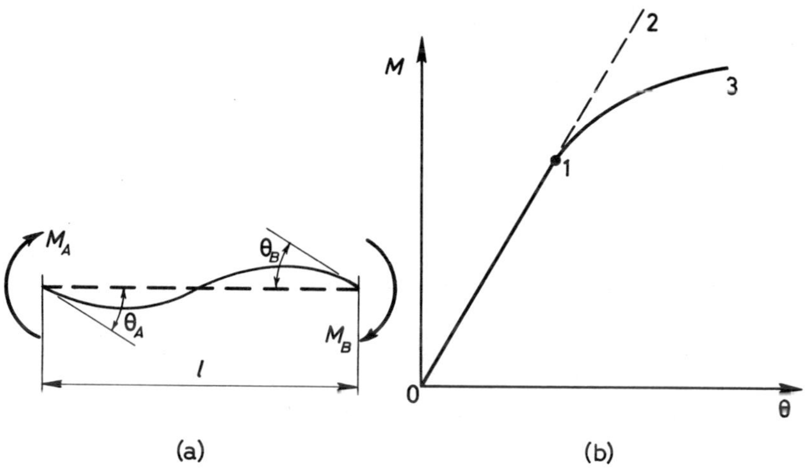

Figure 1.2. Linear and non-linear moment-rotation relationships

superposition is valid in these cases. An exception occurs when the deformations of the structure can no longer be considered *small* and the effect of finite deflections must be taken into account. This occurs in strut theory, where the bending moment induced by an axial load is a function of the lateral deflection of the strut.

The behaviour of a structural element may be linear up to a certain level of load, and thereafter non-linear. This occurs in a steel beam loaded into the plastic or inelastic range. If the beam shown in Fig. 1.2a has a moment rotation relationship, as shown in Fig. 1.2b, the behaviour will be linear between 0 and 1 and non-linear thereafter. In the linear range 0–1 the moment-rotation relationship is

$$\begin{bmatrix} \theta_A \\ \theta_B \end{bmatrix} = \frac{l}{6EI} \begin{bmatrix} 2 & -1 \\ -1 & 2 \end{bmatrix} \begin{bmatrix} M_A \\ M_B \end{bmatrix} \tag{1.4}$$

and in the non-linear range 1–3 the corresponding relationship may be written in the form

$$\begin{bmatrix} \theta_A \\ \theta_B \end{bmatrix} = \frac{l}{6EI}\left\{\begin{bmatrix} 2 & -1 \\ -1 & 2 \end{bmatrix} + \begin{bmatrix} p_A & p_{AB} \\ p_{BA} & p_B \end{bmatrix}\right\}\begin{bmatrix} M_A \\ M_B \end{bmatrix} \tag{1.5}$$

The coefficients p_A, p_B, p_{AB}, and p_{BA} are 'plastic flexibility coefficients', and are functions of the end moments M_A and M_B. They arise from the departure of the line 1–3 from the linear relationship 0–1–2. Non-linear structures are more difficult to solve than linear structures since, in general, direct solutions of the equations are not possible and iterative methods have to be used.[25]

Reciprocal theorem. Consider the propped cantilever shown in Fig. 1.3. When unit vertical load is applied at A the displacement at A is f_{11} and the rotation

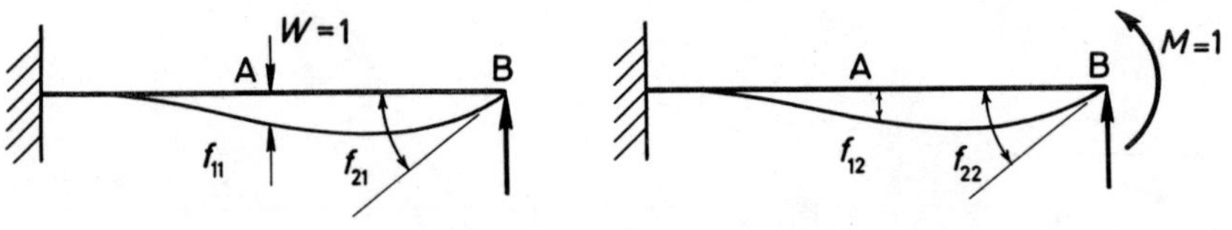

Figure 1.3. Flexibility influence coefficients

at B is f_{21}. When unit moment is applied at B the rotation at B is f_{22} and the vertical displacement at A is f_{12}. If the relationship between load and deformation is linear, the principle of superposition may be used, and the result of applying both forces to the beam will be a rotation at B of $(f_{22} + f_{21})$ and a displacement at A of $(f_{11} + f_{12})$. Clearly, it will be immaterial in which order the forces are applied, the total work done and the final result will be the same. Applying W first then M, the work done is

$$\tfrac{1}{2}Wf_{11} + \tfrac{1}{2}Mf_{22} + Wf_{12} \tag{1.6}$$

Reversing the order and applying M first, followed by W, the work done is

$$\tfrac{1}{2}Mf_{22} + \tfrac{1}{2}Wf_{11} + Mf_{21} \tag{1.7}$$

Equating eqns. (1.6) and (1.7)

$$Wf_{12} = Mf_{21} \tag{1.8}$$

Since the displacements relate to unit values of W and M,

$$f_{12} = f_{21} \tag{1.9}$$

Equation (1.8) is a statement of Maxwell's Reciprocal Theorem for the case of two forces. A more general statement of the theorem for any number of forces is

$$P_1q_1 + P_2q_2 + \cdots + P_nq_n = Q_1p_1 + Q_2p_2 + \cdots + Q_np_n \quad (1.10)$$

in which the set of forces P produces displacements p at corresponding positions and the set of forces Q, at similar positions to the P forces, produces displacements q.

1.2. Equilibrium

Each part of a structure must be in equilibrium under the action of the forces acting on it. The total number of equilibrium conditions will depend on the division of the structure into elements and the type of element employed. It is usually convenient to formulate the equilibrium conditions in terms of forces acting at the nodes or joints in the structure. If A in Fig. 1.4 represents a typical joint in a two-dimensional structure, then for A to be in equilibrium

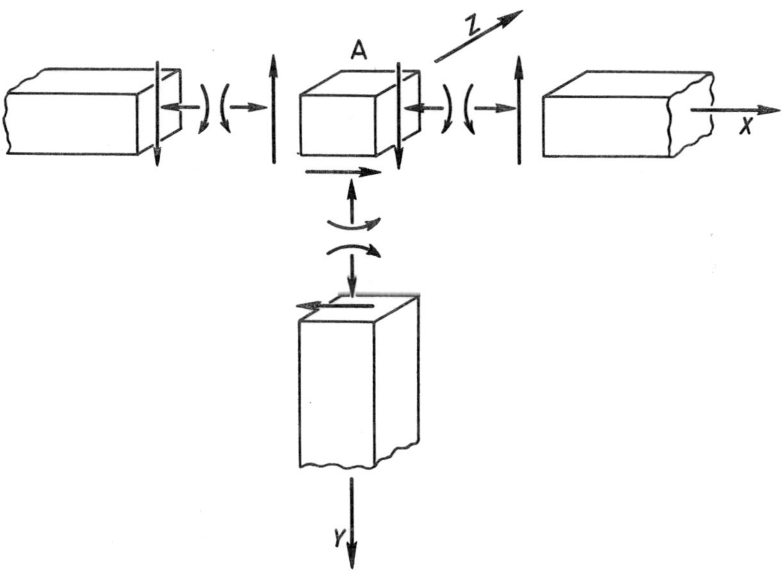

Figure 1.4. Stress resultants in $X - Y$ plane

the sums of the forces (P) acting on A in two perpendicular directions (X, Y) and the sum of the moments (M) in the plane $X - Y$ must be zero, i.e.,

$$\begin{aligned} \sum P_X &= 0 \\ \sum P_Y &= 0 \\ \sum M_{XY} &= 0 \end{aligned} \quad (1.11)$$

In each case, the summation must include any applied loads having components in the X, Y directions and any applied moments in the $X - Y$ plane.

In the analysis of framed structures composed of line elements, no account is taken of the size of joint A and this is considered to be a point in the structure at which the centroidal axes of the members meet. However, the design of the joint itself would require account to be taken of the actual dimensions.

In a three-dimensional structure, such as a space frame, additional forces would have to be considered in formulating the equilibrium conditions. These would be forces in the direction Z, perpendicular to the $X - Y$ plane, moments in the $Y - Z$ plane, and moments in the $Z - X$ plane. Thus six conditions of equilibrium would exist:

$$\begin{aligned} \sum P_X &= 0 \\ \sum P_Y &= 0 \\ \sum P_Z &= 0 \\ \sum M_{XY} &= 0 \\ \sum M_{YZ} &= 0 \\ \sum M_{ZX} &= 0 \end{aligned} \tag{1.12}$$

1.3. Compatibility

The component parts of a structure must deform in a compatible way, i.e., the parts must fit together without discontinuity at all stages in the loading. Consider again the joint A in Fig. 1.4. In the three-dimensional case the joint can displace in the X, Y, Z directions and rotate about the axes X, Y, and Z. These six displacements of the joint must coincide with the displacements of the individual members connected to it. If two of the three members shown in Fig. 1.4 are removed the joint A becomes the end of the remaining member and no compatibility conditions are required. Thus each additional member requires six compatibility conditions and, if there are n rigidly connected members at a joint, the total number of compatibility conditions will be $6(n - 1)$. The corresponding number of compatibility conditions in the two-dimensional case will be $3(n - 1)$.

Stress resultants and releases. The internal actions in numbers such as moment, thrust, and shear result from the stress distribution at the section considered and are termed *stress resultants*. If the structure is designed in such a way that one of these stress resultants is zero for all types of load, then it is said that a *release* has been introduced into the structure. Some examples of releases are shown in Fig. 1.5. A release permits the development of a linear or angular discontinuity between two components in a structure. Each time a release is introduced between two members meeting at a joint the number of compatibility conditions is reduced by one.

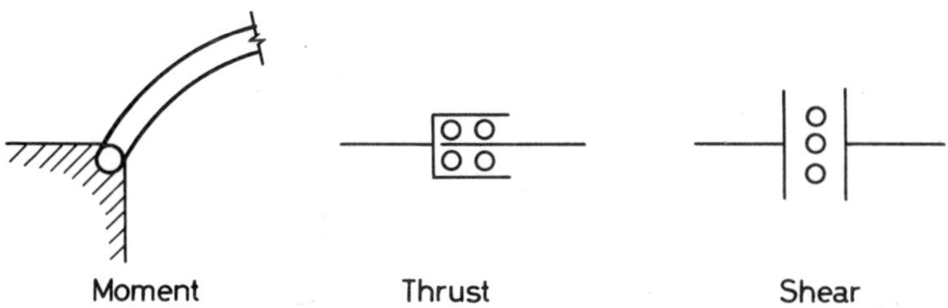

Figure 1.5. Examples of releases

1.4. Static and Kinematic Indeterminacy

If the compatibility conditions for a structure are progressively reduced in number by the introduction of releases, there is reached a state at which the introduction of one further release would convert the structure into a mechanism which, in general, will not support load. In this state, the structure is *statically determinate* and the nodal forces are calculable directly from the equilibrium conditions relating the nodal forces. As the releases are removed and the structure restored to its original condition, further nodal forces will be created and compatibility conditions will be required for their determination. Conversely, a structure is *kinematically determinate* if it is possible to obtain the nodal displacements from compatibility conditions without reference to the equilibrium conditions. A fixed-end beam is kinematically determinate since the end rotations are known from the compatibility conditions of the supports.

Consider the structure shown in Fig. 1.6a, in which the joints A, C, and D are hinged to rigid foundations. The continuity between the foundations through the ground is represented by inserting additional members AD and CD. If the supports A, C, and D are considered encastré, as in Fig. 1.6b, the structure is kinematically determinate except for the displacements of node B. If node B is constrained in such a way that it suffers a rotation only without vertical or horizontal translation, the degree of kinematical indeterminacy,

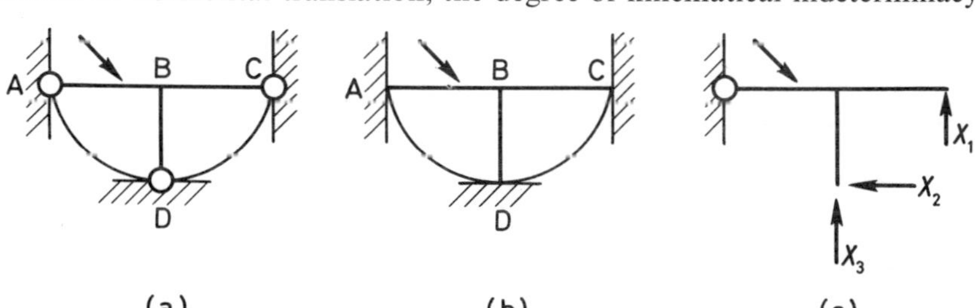

Figure 1.6

n_k, of the structure as shown in Fig. 1.6b is one. The constraints imagined would in fact exist if the members were assumed to have infinitely large extensional rigidities. A *constraint* is thus defined as a device which constrains a displacement at a certain node to be the same as the corresponding displacement, possibly zero, at another node. Thus, if the structure is as shown in Fig. 1.6a, three moment releases have been introduced at A, C, and D. Since the rotations developing at these nodes are not obtainable from compatibility conditions, the degree of kinematical indeterminacy is increased from one to four. This result may be generalized in the expression,

$$n_k = 3(n - 1) - c + r \tag{1.13}$$

in which n is the number of nodes, c the number of constraints, and r the number of releases. In three-dimensional structures, the coefficient of $(n - 1)$ becomes six, since at any typical node there can exist six displacements. The kinematically indeterminate displacements in a structure are known as *degrees of freedom*.

The structure shown in Fig. 1.6b consists of two closed rings when the foundation members AD and CD are included. A complete cut through one of the structural members of each ring, providing it is not the same member, will convert the structure into a statically determinate set of cantilevers. The number of stress resultants released by one cut would be three in this case, six in the case of a three-dimensional structure, thus the degree of statical indeterminacy, n_s, of the structure in Fig. 1.6b is $3 \times$ the number of rings $= 6$. Three releases are introduced in the form of hinges at A, C, and D and thus n_s is reduced to 3 for the structure of Fig. 1.6a. A suitable selection of redundancies $x_{1,2,3}$ is shown in Fig. 1.6c.

The relationship between the number of rings and the numbers of members, M, and nodes, n, can be established by topological principles[16] and is given by

$$\text{number of rings} = {}^{6}_{3}(M - n + 1) \tag{1.14}$$

The coefficient 6 is used for three-dimensional cases and the coefficient 3 for two-dimensional cases. Thus, the expression for the degree of statical indeterminacy is

$$n_k = {}^{6}_{3}(M - n + 1) - r \tag{1.15}$$

Example 1.1. Determine the statical and kinematical indeterminacies of the structures shown in Figs. 1.7a, b, and c.

(a) $n_s = 3(20 - 14 + 1) - 0 = 21$

$n_k = 3(14 - 1) - 3 \times 3 + 0 = 30$

or (assuming zero axial strains in all members),

$n_k = 3(14 - 1) - (3 \times 3 + 10 + 7) + 0 = 13$

(b) $n_s = 6(18 - 12 + 1) - 0 = 42$

$n_k = 6(12 - 1) - 5 \times 6 = 36$

(c) $n_s = 3(7 - 6 + 1) - 4 = 2$
$n_k = 3(6 - 1) - (2 \times 3 + 3) + 4 = 10$
(neglecting axial shortening of columns).

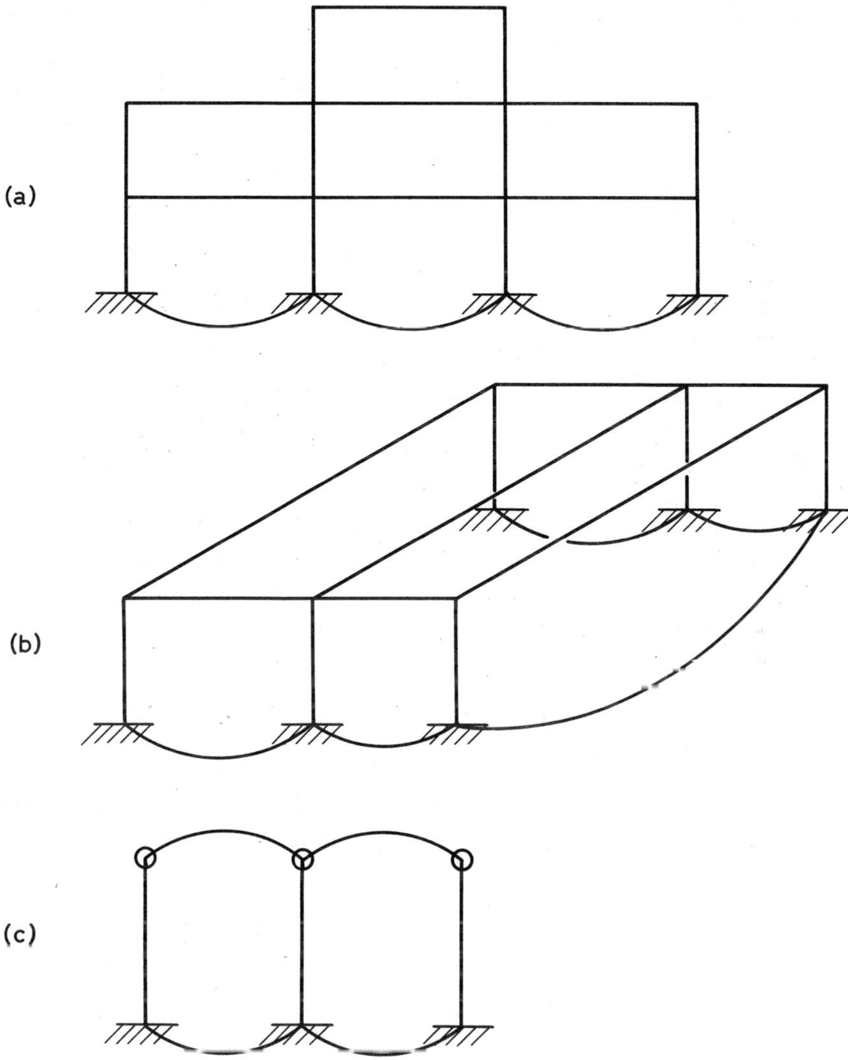

Figure 1.7. Example 1.1

1.5. Particular and Complementary Solutions

We have seen that three basic conditions have to be satisfied in a correct structural analysis: the force-displacement relationships of the elements of the structure; the conditions of equilibrium; and the conditions of compatibility. Two parallel courses of action are possible. Observing the force-displacement requirements throughout we may: (a) establish equilibrium in any arbitrary manner and superimpose a solution to satisfy compatibility, or (b) establish continuity, again in any convenient way, and superimpose a solution to restore equilibrium. In both cases, the first solution is termed a *particular solution* and the second a *complementary solution*. Thus, in either case,

$$\text{Total solution} = \text{particular solution} + \text{complementary solution} \quad (1.16)$$

If we follow method (a), we render the structure statically determinate by introducing releases and allotting arbitrary, usually zero, values to the corresponding stress resultants, *redundancies*. The resulting structure is known as the *primary structure*. On application of load to the primary structure, discontinuities will develop at the releases. The discontinuities are calculated and the complementary solution developed to restore compatibility at the releases. The values of the redundants are obtained from the conditions that the discontinuities they produce in the primary structure must be the same, but of opposite sign to those produced by the applied loads. The method is called the *force* or *flexibility* method, the unknowns in the solution being the forces developed at the releases necessary to restore compatibility.

If method (b) is followed specific, usually zero, values are allotted to the nodal degrees of freedom of the structure. Thus, additional constraints are added to the structure to produce the primary structure. The complementary solution in this case restores equilibrium by removing the unwanted constraining forces. This is the *displacement* or *stiffness* method, the unknowns being the nodal displacements in the complementary solution required to restore equilibrium. The following example, Fig. 1.8, will serve to illustrate the two methods.

In this example, a single compatibility or equilibrium condition is sufficient to obtain the complementary solution. In general, we shall have more than one unknown and the equations

$$\mathbf{fx} + \mathbf{u} = \mathbf{0} \quad (1.17)$$

and

$$\mathbf{kr} + \mathbf{R} = \mathbf{0} \quad (1.18)$$

will be matrix equations with the matrices **f** and **k**, more usually **F** and **K**, being square matrices of order equal to the number of unknowns.

1.6. Flexibility

The concept of an influence coefficient was introduced in eqns. (1.1) and (1.2) and has been used again in the example in Fig. 1.8, in which f and k are respectively *flexibility* and *stiffness* influence coefficients. A flexibility influence coefficient, f_{rs}, is defined as the displacement in direction r in a structure due to unit force in direction s. If the directions in which we are interested are numbered 1 to n, and r and s take successive values in the range

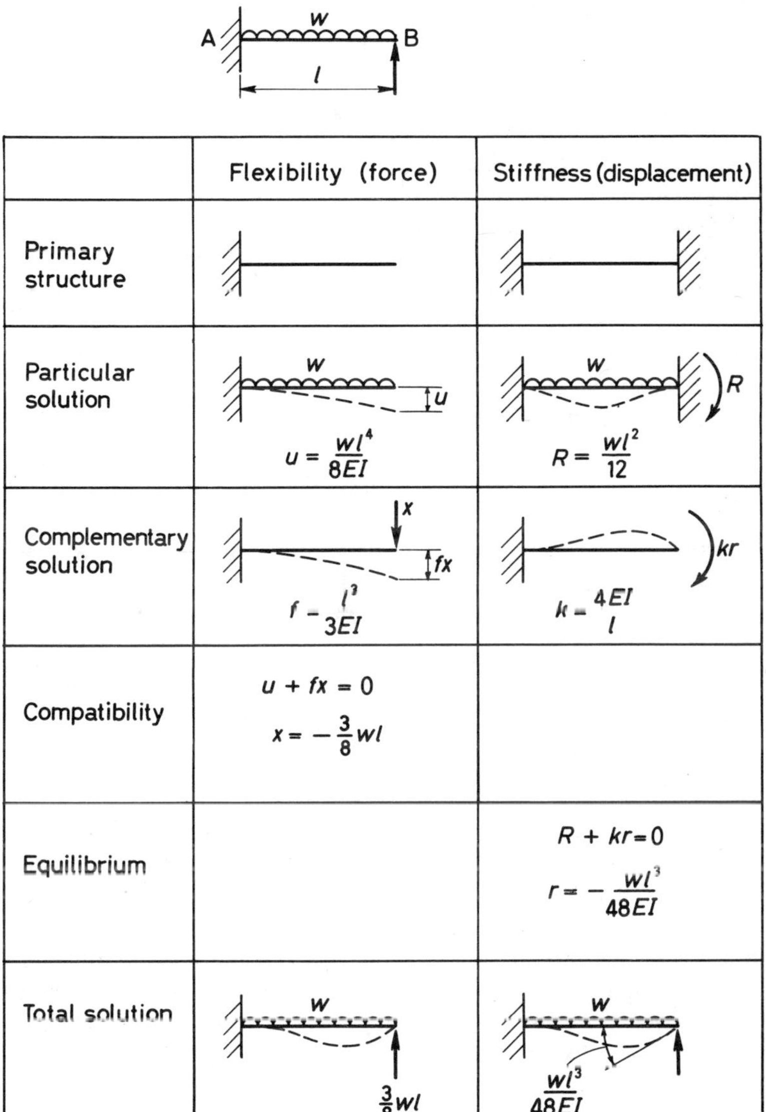

Figure 1.8. Example of alternative analyses by the flexibility and stiffness methods

$1 \ldots n$, there will be n^2 flexibility influence coefficients. It is unnecessary, however, to calculate all the n^2 coefficients since, as we saw in section 1.1, $f_{rs} = f_{sr}$ from the reciprocal theorem.

Figure 1.9 shows a cantilever beam in which three directions have been chosen for which influence coefficients are required. Unit values of forces are applied successively in the three directions and the resulting displacements are the flexibility influence coefficients. It should be noted that although f_{32} and f_{23} are numerically equal in this case, f_{32} is an angular displacement and f_{23} a linear displacement. This is perhaps seen more clearly if the consistency of units in eq. (1.8) is studied.

Calculation of flexibility influence coefficients. The process of calculating flexibility influence coefficients is essentially that of calculating displacements at discrete points in the structure. For simplicity, we shall consider a structure in which bending deformations predominate. The extension of the procedure to include other causes of deformation is straightforward and will be discussed in chapter 3.

Referring to the example in Fig. 1.9, it is required to obtain the nine influence coefficients corresponding to directions 1, 2, and 3. This is best done by using Castigliano's theorem,

$$\Delta_i = \int M \, \partial M/\partial F_i \frac{ds}{EI} \tag{1.19}$$

in which the integration is taken over the whole structure. M is the bending moment distribution due to the applied loads, and thus $\partial M/\partial F_i$ can be regarded as the bending moment distribution caused by unit force at position i where the deflection is required. The bending moment distributions due to unit value of each of the three forces in turn are shown in Fig. 1.9 labelled m_1, m_2, and m_3. Thus, it follows from eq. (1.19) that the influence coefficients are

$$f_{rs} = \int m_r m_s \frac{ds}{EI} \tag{1.20}$$

The evaluation of an influence coefficient thus requires the integration of a product of two bending moment distributions over the complete structure. The diagrams are often very simple combinations of rectangles, triangles, parabolas, and trapezia and it is convenient to establish standard results for such product integrals in advance. Appendix A shows product integrals for the more frequently occurring combinations of bending moment diagrams. It should be noted that the flexural rigidity EI occurs in the denominator of the integral in eq. (1.20) and the results quoted in Appendix A assume EI constant over the length of the diagram. If EI varies the integrals have to be evaluated by other methods. This type of calculation will be considered in chapter 3.

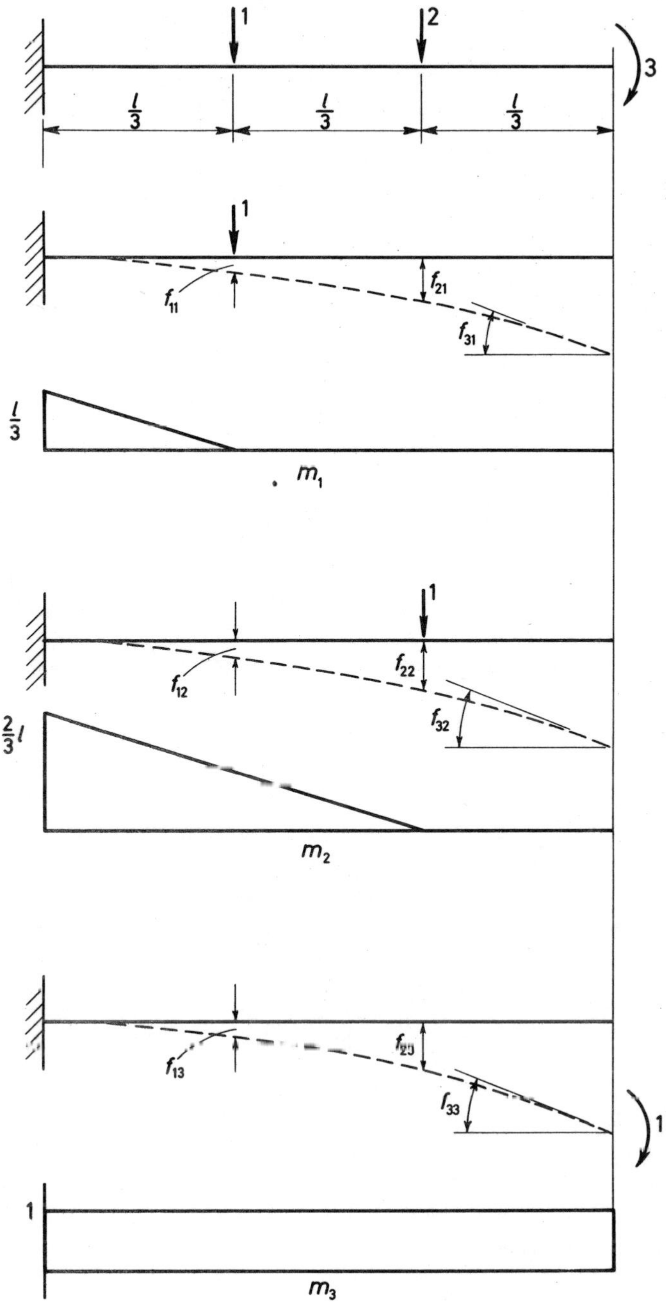

Figure 1.9. Flexibility influence coefficients

Referring to the example in Fig. 1.9, we can now obtain the influence coefficients as follows:

$$f_{11} = \int m_1{}^2 \frac{ds}{EI} = \frac{1}{3}\cdot\frac{l}{3}\cdot\frac{l}{3}\cdot\frac{l}{3}\cdot\frac{1}{EI} = \frac{l^3}{81EI}$$

$$f_{22} = \int m_2{}^2 \frac{ds}{EI} = \frac{1}{3}\cdot\frac{2l}{3}\cdot\frac{2l}{3}\cdot\frac{2l}{3}\cdot\frac{1}{EI} = \frac{8l^3}{81EI}$$

$$f_{33} = \int m_3{}^2 \frac{ds}{EI} = \frac{l.1.1}{EI} = \frac{l}{EI}$$

$$f_{12} = f_{21} = \int m_1 m_2 \frac{ds}{EI} = \frac{1}{6}\cdot\frac{l}{3}\cdot\frac{l}{3}\left(\frac{4l}{3} + \frac{l}{3}\right)\frac{1}{EI} = \frac{5l^3}{162EI}$$

$$f_{13} = f_{31} = \int m_1 m_3 \frac{ds}{EI} = \frac{1}{2}\cdot\frac{l}{3}\cdot\frac{l}{3}\cdot\frac{1}{EI} = \frac{l^2}{18EI}$$

$$f_{23} = f_{32} = \int m_2 m_3 \frac{ds}{EI} = \frac{1}{2}\cdot\frac{2l}{3}\cdot\frac{2l}{3}\cdot\frac{1}{EI} = \frac{2l^2}{9EI}$$

Effect of applied loads on primary structure. When loads are applied to the primary structure discontinuities develop at the releases. In the example in Fig. 1.8, the applied load produces a vertical displacement at B,

$$u = \frac{wl^4}{8EI}$$

This displacement is reduced to zero, to satisfy the condition of compatibility at support B, by the correct value of the redundant force, $x = -\frac{3}{8}wl$. The calculation of the displacement u is carried out in a similar way to the calculation of the flexibility influence coefficient f. If u_i is the displacement in the primary structure at the position, and in the direction, of the ith redundancy caused by the applied loads, then

$$u_i = \int m_0 m_i \frac{ds}{EI} \tag{1.21}$$

where m_0 is the bending moment distribution in the primary structure due to the applied loads. In the calculation of the u_i ($i = 1, 2, \ldots, n_s$) the table of product integrals can be used, providing EI is constant, in the same way as with the f_{rs}.

Compatibility conditions. In the example of Fig. 1.8, a single compatibility condition sufficed to effect a solution and only one equation had to be solved. In general, the compatibility equations will take the matrix form

$$\mathbf{FX} + \mathbf{U} = \mathbf{0} \tag{1.22}$$

and the solution for the vector **X** requires the solution of a set of linear simultaneous equations of which the coefficient matrix is **F**.

Sign conventions. The bending moment expressions used in evaluating influence coefficients from eqns. (1.20) and (1.21) must carry signs based on the type of curvature developing in the beam. Since the integrand is always the product of two bending moment expressions, it is only the relative sign that is of importance. If the bending moment diagrams are sketched on the convex, or tension, side of the member the relative signs of m_r and m_s can easily be seen. If the two diagrams are on the same side of the member the product $m_r m_s$ is positive, and if drawn on opposite sides of the member the product is negative.

Example 1.2. The structure shown in Fig. 1.10 is a propped cantilever with an additional support at mid-span in the form of a spring. The flexibility of the spring is assumed to be $l^3/24EI$, where EI is the uniform flexural rigidity of the beam. It is required to obtain the bending moment distribution in the beam due to an overall uniformly distributed load.

$$f_{11} = \int {m_1}^2 \frac{ds}{EI} = \frac{2}{3} l \cdot 1 \cdot \frac{1}{EI} = \frac{2l}{3EI}$$

$$f_{22} = \int {m_2}^2 \frac{ds}{EI} = 2 \cdot \frac{l}{3} \cdot \frac{l}{2} \cdot \frac{l}{2EI} = \frac{l^3}{6EI}$$

$$f_{12} = f_{21} = \int m_1 m_2 \frac{ds}{EI} = \frac{l}{6} \cdot \frac{l}{2} \left(\frac{1+1}{EI} \right) + \frac{l}{3} \cdot \frac{l}{2} \cdot \frac{1}{2} \cdot \frac{1}{EI} = \frac{l^2}{4EI}$$

$$u_1 = \int m_0 m_1 \frac{ds}{EI} = -\frac{2l}{3} \cdot 1 \cdot \frac{wl^2}{2EI} = -\frac{wl^3}{3EI}$$

$$u_2 = \int m_0 m_2 \frac{ds}{EI} = -\frac{5}{12} \cdot \frac{2l}{EI} \cdot \frac{l}{2} \cdot \frac{wl^2}{2} = -\frac{5}{24} \frac{wl^4}{EI}$$

The compatibility condition at x_1 is

$$f_{11} x_1 + f_{12} x_2 + u_1 = 0$$

and at x_2, (1.23)

$$f_{21} x_1 + f_{22} x_2 + u_2 = -x_2 \frac{l^3}{24EI}$$

that is,

$$\frac{2l}{3} x_1 + \frac{l^2}{4} x_2 = \frac{wl^3}{3}$$

(1.24)

and

$$\frac{l^2}{4} x_1 + \left(\frac{l^3}{6} + \frac{l^3}{24} \right) x_2 = \frac{5}{24} wl^4$$

from which
$$x_1 = \frac{5}{22} wl^2$$
$$x_2 = \frac{8}{11} wl \tag{1.25}$$

Finally, the bending moment at any section of the beam is obtained by addition of the bending moments due to the applied loads and the redundancies,

$$M = m_0 + m_1 x_1 + m_2 x_2 \tag{1.26}$$

giving the bending moment diagram shown in Fig. 1.10.

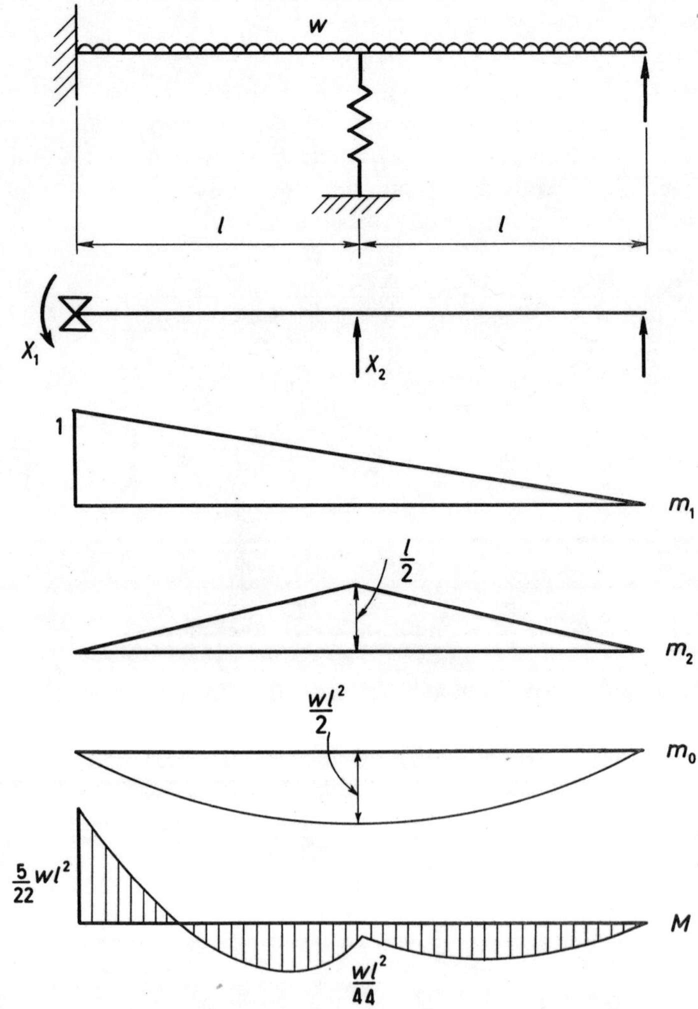

Figure 1.10. Example 1.2

1.7. Stiffness

If a structure has n degrees of freedom of nodal displacements and if these displacements are given unit value in turn, while the remaining nodal displacements are constrained, certain forces will be required at the constraints. A stiffness influence coefficient k_{rs} is defined as the force required at constraint r when unit value is given to displacement s. The subscripts r and s will take, successively, values in the range $1 \ldots n$, and thus there will be n^2 stiffness influence coefficients. When $r = s$, the force developed acts at the same position and in the same direction as the unit displacement and the stiffness is termed a *direct stiffness*. When $r \neq s$ the force developed is at some other position or direction from the unit displacement and the stiffness is termed a *cross stiffness*.

Figure 1.11 shows the direct and cross stiffnesses for a uniform beam with two degrees of freedom at each end. In the case of a straight, uniform-section beam, the stiffnesses k_{rs} are well known and the stiffness matrix $\mathbf{k}$ can be written down by inspection.

$$\mathbf{k} = \begin{bmatrix} k_{11} & k_{12} & k_{13} & k_{14} \\ k_{21} & k_{22} & k_{23} & k_{24} \\ k_{31} & k_{32} & k_{33} & k_{34} \\ k_{41} & k_{42} & k_{43} & k_{44} \end{bmatrix} = EI \left[\begin{array}{cc|cc} 4/l & 6/l^2 & 2/l & -6/l^2 \\ 6/l^2 & 12/l^3 & 6/l^2 & -12/l^3 \\ \hline 2/l & 6/l^2 & 4/l & -6/l^2 \\ -6/l^2 & -12/l^3 & -6/l^2 & 12/l^3 \end{array}\right]$$

$$= EI \begin{bmatrix} k_A & k_{AB} \\ k_{AB}^T & k_B \end{bmatrix} \quad (1.27)$$

The stiffnesses in eqns. (1.27) form a symmetric matrix, i.e., one in which $k_{rs} = k_{sr}$. This property follows from the reciprocal theorem. For example, if we apply the reciprocal theorem, eq. (1.10) to b and c of Fig. 1.11,

$$k_{11}(0) + k_{21}(1) + k_{31}(0) + k_{41}(0) = k_{12}(1) + k_{22}(0) + k_{32}(0) + k_{42}(0)$$

thus, $$k_{21} = k_{12} \quad (1.28)$$

It is sometimes convenient to partition the stiffness matrix as in the last of eqns. (1.27), in which the submatrices k_A and k_B contain the direct and cross stiffnesses relating to ends A and B respectively whereas the submatrix k_{AB} contains the cross stiffnesses relating one end to the other.

Calculation of stiffness influence coefficients. In structures composed of straight, uniform, members in which axial strains are considered negligible, the stiffness matrix in eqns. (1.27) contains all the stiffnesses required for the majority of practical applications. In cases where the member is of non-uniform flexural rigidity EI, or where deformations other than those due to bending have to be considered, the stiffnesses can be obtained from the end flexibilities using the inverse relationship between stiffness and flexibility given in eq. (1.3). A general method of calculating stiffnesses of non-uniform section members will be considered in chapter 5 and the development of stiffness matrices for more complex elements will be discussed in chapter 6.

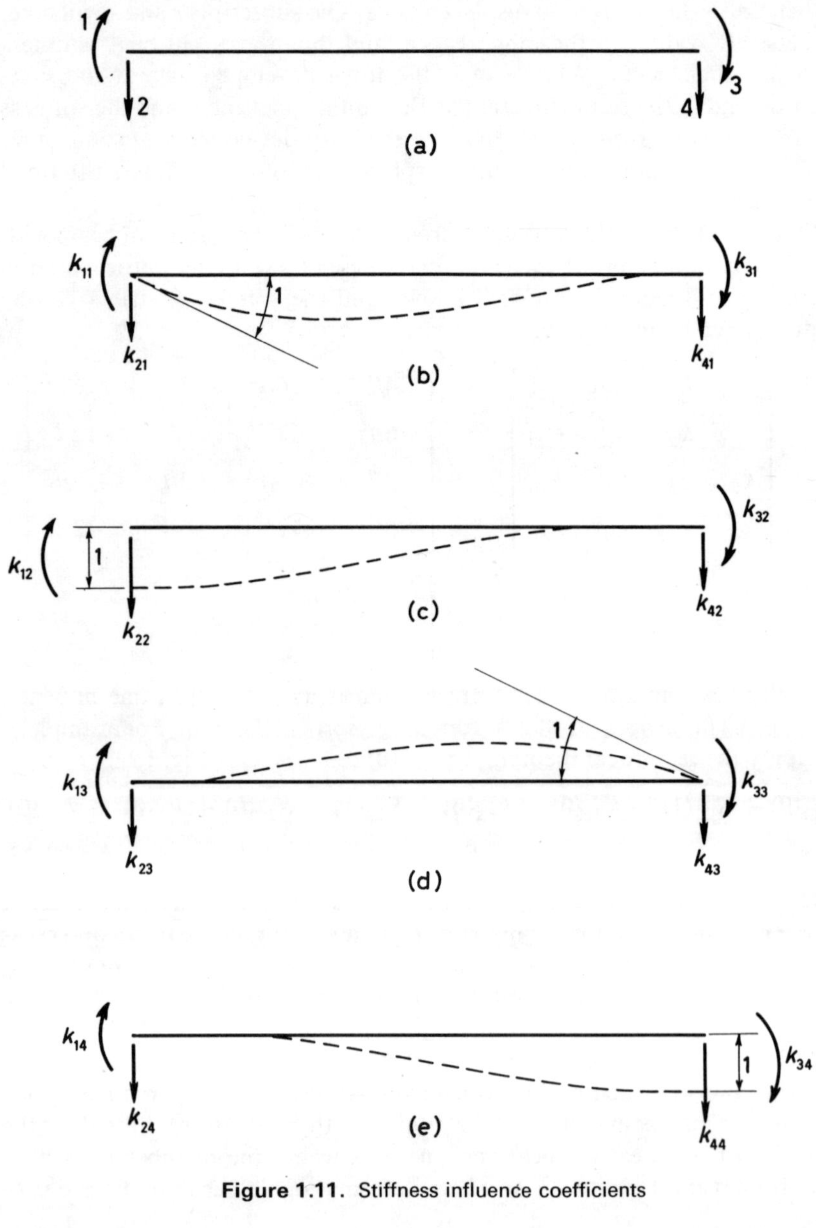

Figure 1.11. Stiffness influence coefficients

Effect of applied loads on primary structure. The primary structure in the stiffness method is one in which the degrees of freedom of nodal displacements are constrained. The usual procedure is to apply constraints completely inhibiting all the degrees of freedom, but there are certain situations in which it is convenient to constrain a particular degree of freedom at a certain, non-zero, value of displacement. When external loads are applied to the primary structure, some or all of the constraints will be required to exert forces to prevent the development of nodal displacements. The constraining forces induce systems of fixed-end forces and moments in the members of the structure constituting the particular solution. The constraining forces are removed by developing a complementary solution in which the applied loads are taken to be the same as the constraining forces, but with all the signs reversed. Thus, the superposition of the particular and complementary solutions removes the lack of equilibrium caused by the constraining forces.

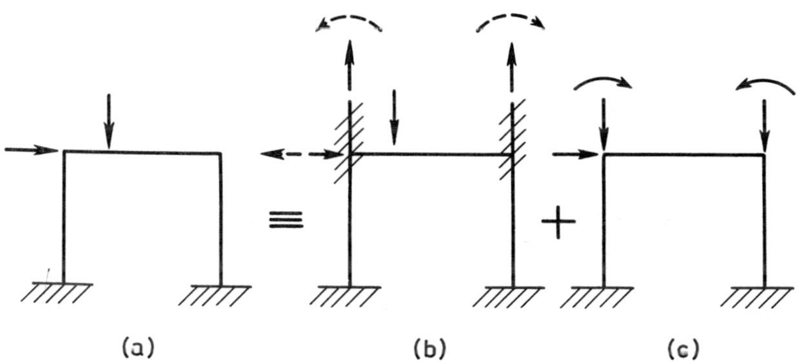

Figure 1.12. (a) Applied loading system; (b) loading system for particular solution; (c) loading system for complementary solution

In the structure shown in Fig. 1.12, (a) shows the applied loads, (b) the constrained structure carrying the applied loads with the constraining forces dotted, and (c) shows the constraining forces, reversed in sign, as applied to the structure for the determination of the complementary solution.

Equilibrium conditions. The nodal displacements accompanying the application of external forces must be such as to induce an internal force distribution with nodal stress resultants in equilibrium with the applied loads. Thus, for every nodal degree of freedom there will correspond an equilibrium condition. For example, if we consider the continuous beam shown in Fig. 1.13, there are four degrees of freedom of rotation at the supports numbered $r_{1...4}$. The stiffness influence coefficients are found by prescribing unit value to each rotation in turn. The final values of the joint rotations must be such as to achieve equilibrium between internal and external forces. The internal moment developed at support 1 is

$$k_{11}r_1 + k_{12}r_2$$

the rotations r_3 and r_4 have no effect since the corresponding cross stiffnesses k_{13} and k_{14} are zero. Thus, for equilibrium at support 1,

$$k_{11}r_1 + k_{12}r_2 \qquad\qquad = R_1$$

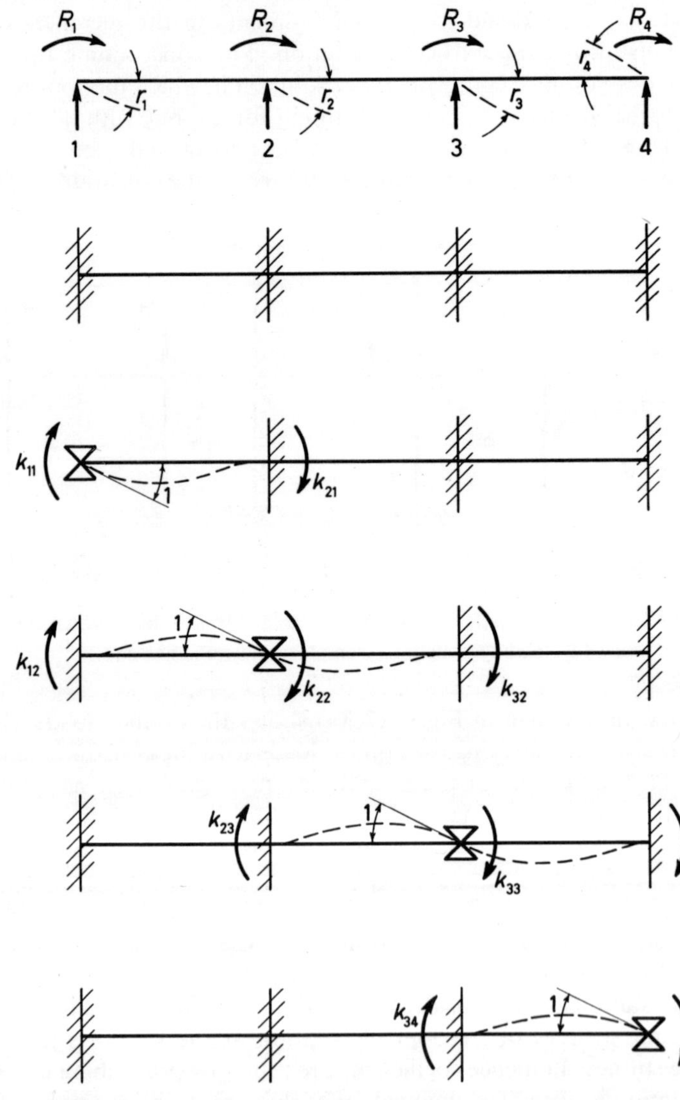

Figure 1.13

and similarly at supports 2, 3, and 4,

$$\begin{aligned} k_{21}r_1 + k_{22}r_2 + k_{23}r_3 \qquad\qquad &= R_2 \\ k_{32}r_2 + k_{33}r_3 + k_{34}r_4 &= R_3 \\ k_{43}r_3 + k_{44}r_4 &= R_4 \end{aligned} \tag{1.29}$$

In matrix form, the eqns. (1.29) are

$$\mathbf{Kr} = \mathbf{R} \tag{1.30}$$

in which

$$\mathbf{K} = \begin{bmatrix} k_{11} & k_{12} & & \\ k_{21} & k_{22} & k_{23} & \\ & k_{32} & k_{33} & k_{34} \\ & & k_{43} & k_{44} \end{bmatrix}; \quad \mathbf{r} = \begin{bmatrix} r_1 \\ r_2 \\ r_3 \\ r_4 \end{bmatrix}; \quad \mathbf{R} = \begin{bmatrix} R_1 \\ R_2 \\ R_3 \\ R_4 \end{bmatrix} \tag{1.31}$$

Sign convention. The nodal displacements are numbered and positive directions are assumed. The signs of the applied loads constituting the vector **R** in eqns. (1.31) are then positive or negative, depending on whether they are in the same or reverse sense as the nodal displacements. If the subsequent calculation reveals negative signs for one or more of the nodal displacements, these are then in opposite sense to the assumed directions.

Example 1.3. To illustrate the method, and at the same time provide a comparison with the flexibility method, example 1.2 will now be solved using the stiffness method.

The degree of kinematical indeterminacy, using eq. (1.13), is

$$n_k = 3(3 - 1) - 6 + 3 = 3$$

These are the rotations of the beam at B and C and the vertical displacement at B, Fig. 1.14. The stiffness influence coefficients are shown in Fig. 1.14, and these can be written down by inspection using the standard results from the stiffness matrix of eq. (1.27). The spring support at B has a given flexibility, i.e., displacement due to unit force, of $l^3/24EI$. Its stiffness, i.e., force to produce unit displacement, is the reciprocal of this, namely $24EI/l^3$, thus

$$k_{11} = 8EI/l; \quad k_{22} = 24EI/l^3 + 24EI/l^3; \quad k_{33} = 4EI/l$$

$$k_{12} = k_{21} = 0; \quad k_{13} = k_{31} = 2EI/l; \quad k_{23} = k_{32} = 6EI/l^2$$

$$R_1 = wl^2/12 - wl^2/12 = 0; \quad R_2 = wl/2 + wl/2 = wl;$$

$$R_3 = -wl^2/12.$$

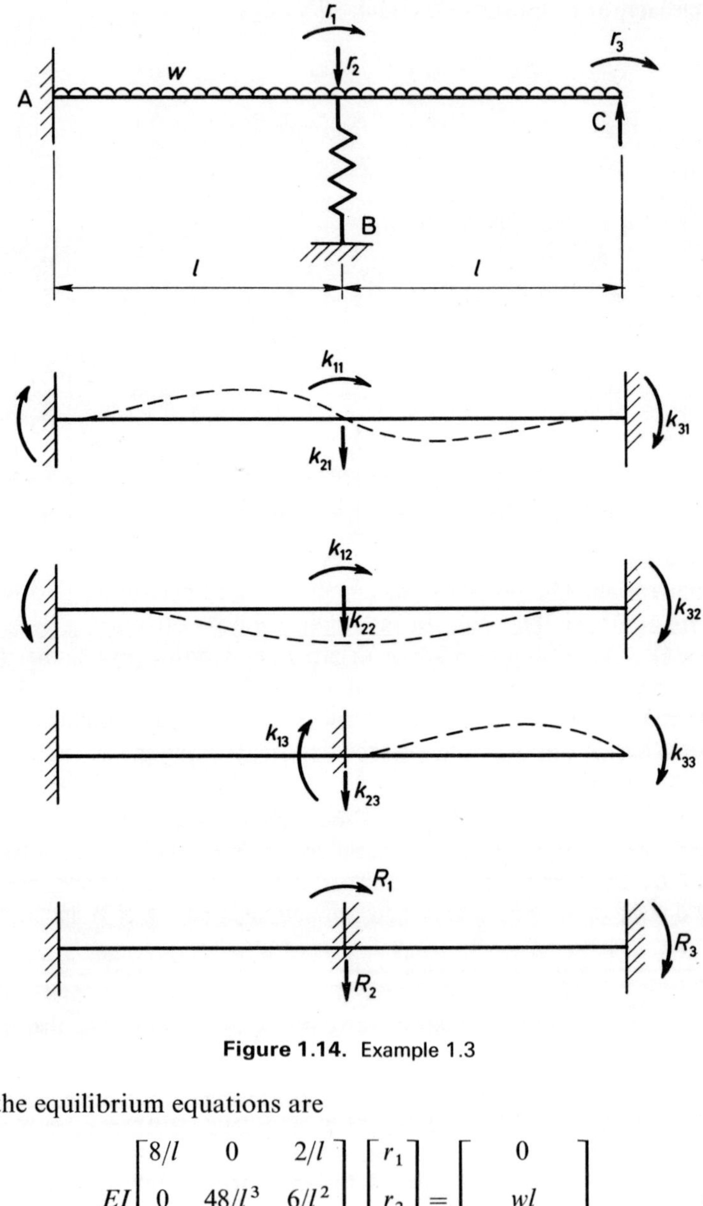

Figure 1.14. Example 1.3

Thus, the equilibrium equations are

$$EI\begin{bmatrix} 8/l & 0 & 2/l \\ 0 & 48/l^3 & 6/l^2 \\ 2/l & 6/l^2 & 4/l \end{bmatrix}\begin{bmatrix} r_1 \\ r_2 \\ r_3 \end{bmatrix} = \begin{bmatrix} 0 \\ wl \\ -wl^2/12 \end{bmatrix} \tag{1.32}$$

The solutions of eqns. (1.32) are

$$r_1 = \frac{5}{264}\frac{wl^3}{EI}; \qquad r_2 = \frac{wl^4}{33EI}; \qquad r_3 = -\frac{5}{66}\frac{wl^3}{EI}$$

The reaction from the spring is

$$\frac{24EI}{l^3} r_2 = \frac{8}{11} wl$$

The bending moments in the two spans are obtained by multiplying the displacements by the appropriate stiffness coefficients from eq. (1.27). For example,

$$\begin{aligned} M_A &= 4EI/l(0) + 6EI/l^2(0) + 2EI/l(r_1) - 6EI/l^2(r_2) \\ &= -\frac{19}{132} wl^2 \end{aligned}$$

To this moment must be added the fixing moment from the particular solution, thus,

$$M_A = -\frac{19}{132} wl^2 - \frac{wl^2}{12} = -\frac{5}{22} wl^2$$

Relationship between flexibility and stiffness matrices. It was stated in section 1.1 that the flexibility matrix is the inverse of the stiffness matrix. This is true only when both matrices relate to the same structure and to corresponding positions and directions in the structure. The flexibility matrix developed in example 1.2, eqns. (1.23), is not the inverse of the stiffness matrix, eqns. (1.32) in example 1.3. The flexibility matrix is that for the primary structure and is not the flexibility matrix of the actual structure, which would be the inverse of the stiffness matrix.

Assembly of stiffness matrix. In the previous example, the assembly of the stiffness matrix was accomplished by inspection, referring to a diagram of the structure and the system of numbering the nodal displacements. When the stiffness method is applied to a digital computer the process of matrix assembly must be made automatic. This involves the description, in some convenient terms, of the geometry and layout of the structure, in such a way that it can be interpreted by the computer in much the same way as we would use a sketch diagram of the structure. Methods of automatic assembly of stiffness matrices will be considered in chapter 4, but a useful concept to introduce at this stage is that of matrix transformation.

1.8. Matrix Transformation of Coordinates

No attempt will be made here to deal exhaustively with the subject of matrix transformations. For this, the reader is referred to any standard text on matrix theory.[11, 30] From the point of view of structural analysis, matrix transformation permits an algebraic, as distinct from geometrical, transfer of effects, such as displacement and force, from one set of coordinate axes to another. It was pointed out in the introduction that it is convenient to regard a structure as an assemblage of elements for which force-displacement relationships are known. The coordinate system used in describing the element force-displacement relationships will be one conveniently related to the geometry of the element; for example, the origin of the coordinate system for a beam element will usually be taken at one end, and one of the coordinate axes directed along the axis of the beam. The forces and displacements will be listed in some order which has to be related to the ordering of forces and displacements in the actual structure. When structural computations are done by hand, a visual inspection process plays a large part in relating element properties to structural properties. Matrix transformation theory permits the transfer of information from element to structure in a completely algebraic way, and thus forms the basis of an automatic process suitable for digital computers.

Suppose a set of forces $A_{1,2,3}$ and corresponding displacements $a_{1,2,3}$ exist at point 0 in a structure, and we wish to find the equivalent effects $B_{1,2,3}$ and $b_{1,2,3}$ at point 0′, having coordinates (x, y) with 0 as origin, Fig. 1.15. The relationships between the two sets of forces are obtained from statical considerations, thus,

$$\begin{aligned} B_1 &= A_1 \\ B_2 &= A_2 \\ B_3 &= A_3 + A_2 x - A_1 y \end{aligned} \quad \text{i.e.,} \quad \begin{bmatrix} B_1 \\ B_2 \\ B_3 \end{bmatrix} = \begin{bmatrix} 1 & 0 & 0 \\ 0 & 1 & 0 \\ -y & x & 1 \end{bmatrix} \begin{bmatrix} A_1 \\ A_2 \\ A_3 \end{bmatrix} \tag{1.33}$$

Assuming that the displacements are small, within the usual structural meaning of the term, then the two systems of displacements are related as follows:

$$\begin{aligned} b_1 &= a_1 + a_3 y \\ b_2 &= a_2 - a_3 x \\ b_3 &= a_3 \end{aligned} \quad \text{i.e.,} \quad \begin{bmatrix} b_1 \\ b_2 \\ b_3 \end{bmatrix} = \begin{bmatrix} 1 & 0 & y \\ 0 & 1 & -x \\ 0 & 0 & 1 \end{bmatrix} \begin{bmatrix} a_1 \\ a_2 \\ a_3 \end{bmatrix} \tag{1.34}$$

Equations (1.33) and (1.34) may be written in the matrix form:

$$\mathbf{B} = \mathbf{T}_1 \mathbf{A} \tag{1.35}$$

$$\mathbf{b} = \mathbf{T}_2 \mathbf{a} \tag{1.36}$$

If either of the transformation matrices $\mathbf{T}_1$, $\mathbf{T}_2$ is postmultiplied by the transpose of the other, the product is seen to be the unit matrix.

$$\begin{bmatrix} 1 & 0 & 0 \\ 0 & 1 & 0 \\ -y & x & 1 \end{bmatrix} \begin{bmatrix} 1 & 0 & 0 \\ 0 & 1 & 0 \\ y & -x & 1 \end{bmatrix} = \begin{bmatrix} 1 & 0 & 0 \\ 0 & 1 & 0 \\ 0 & 0 & 1 \end{bmatrix}$$

that is,

$$\mathbf{T}_1 \mathbf{T}_2^T = \mathbf{I} \tag{1.37}$$

or $$\mathbf{T}_2\mathbf{T}_1{}^T = \mathbf{I} \tag{1.38}$$

Hence $$\mathbf{T}_1 = (\mathbf{T}_2{}^T)^{-1}$$

or $$\mathbf{T}_2 = (\mathbf{T}_1{}^T)^{-1}$$

Thus, in eq. (1.36), $$\mathbf{b} = (\mathbf{T}_1{}^T)^{-1}\mathbf{a}$$

or $$\mathbf{a} = \mathbf{T}_1{}^T\mathbf{b} \tag{1.39}$$

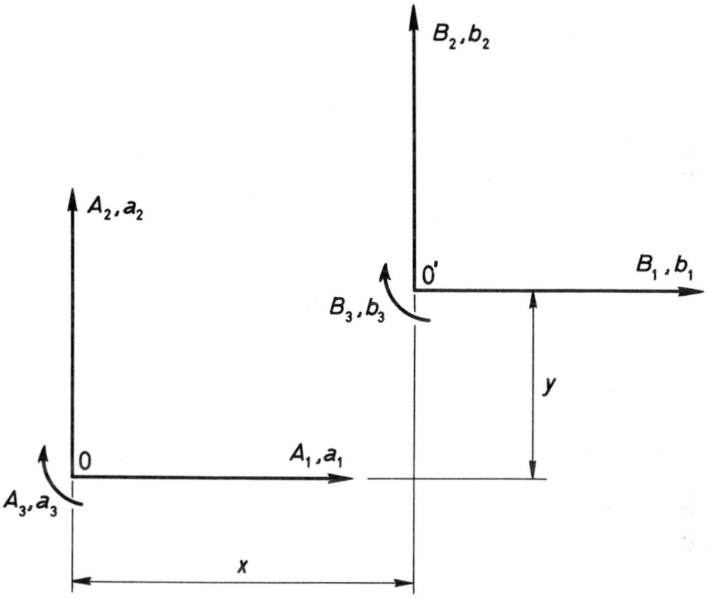

Figure 1.15. Axis translation

The relationships $\mathbf{B} = \mathbf{T}_1\mathbf{A}$ and $\mathbf{a} = \mathbf{T}_1{}^T\mathbf{b}$ illustrate the principle of *contragredience* of force and displacement. It is said that forces and displacements transform *contragrediently*. This principle will be useful in chapters 2 and 4 when the flexibility and stiffness methods are developed completely in matrix terms.

The principle of contragredience may also be established using virtual work. If virtual displacements $\delta\mathbf{a}$ are imposed on system A at 0, then the corresponding virtual displacements $\delta\mathbf{b}$ at 0′ must be such that the work done is the same,

that is, $$A_1\,\delta a_1 + A_2\,\delta a_2 + A_3\,\delta a_3 = B_1\,\delta b_1 + B_2\,\delta b_2 + B_3\,\delta b_3$$

or $$\mathbf{A}^T\,\delta\mathbf{a} = \mathbf{B}^T\,\delta\mathbf{b}$$

$$= \mathbf{A}^T\mathbf{T}_1{}^T\,\delta\mathbf{b}$$

from eq. (1.35),

hence $$\delta\mathbf{a} = \mathbf{T}_1{}^T\,\delta\mathbf{b}$$

Since the displacements are assumed to be small, then $\delta\mathbf{a}$ and $\delta\mathbf{b}$ may be replaced by real displacements **a** and **b**, thus,

$$\mathbf{a} = \mathbf{T}_1^{T}\mathbf{b}$$

as in eq. (1.39).

PROBLEMS

1.1. The three-span continuous beam of uniform section shown in Fig. 1.16 carries a load W on the centre span, distances a from support B and b from support C.

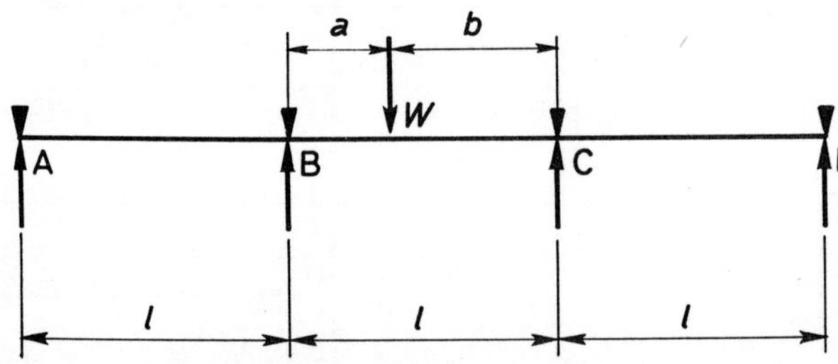

Figure 1.16. Problem 1.1

Choosing x_1 and x_2 as redundants, the internal bending moments in the beam at B and C respectively, assemble a primary structure flexibility matrix and calculate M_B, M_C and the bending moment under the load W.

1.2. Analyse the portal frame shown in Fig. 1.17 by the flexibility method, and draw the bending moment diagram due to the loading shown. The column feet are fixed at A and hinged at D, and the beam has a flexural rigidity which is twice that of the columns.

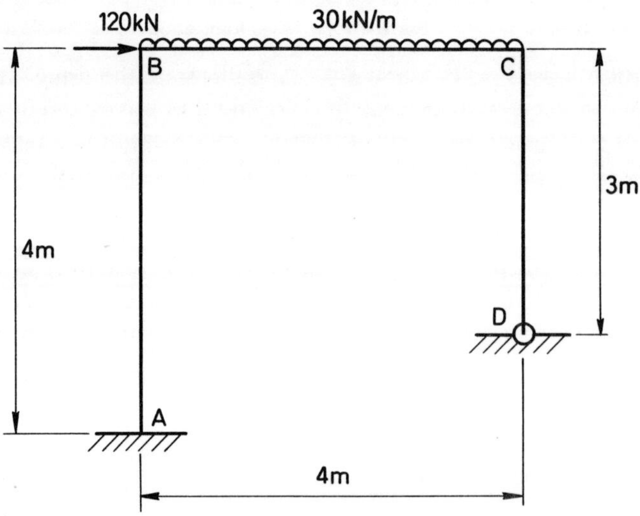

Figure 1.17. Problem 1.2

1.3. The beams in the two-storey frame in Fig. 1.18 have a cross-section such that the flexural rigidity may be considered infinitely large in comparison with that of the

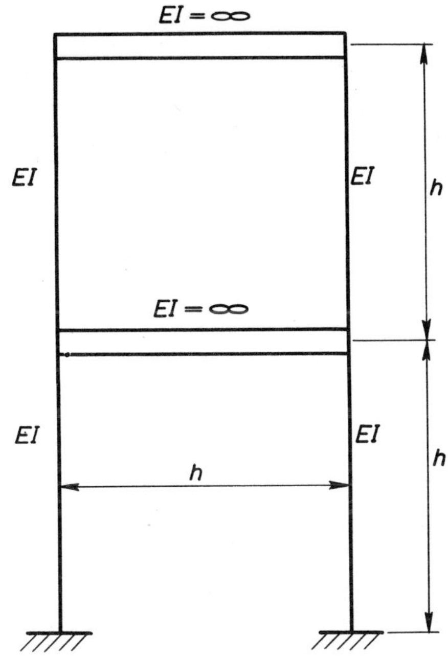

Figure 1.18. Problem 1.3

columns. Both columns and beams may be considered to have infinitely large extensional rigidities. Assemble a stiffness matrix **K** for the structure and also a flexibility matrix **F**. Show that $\mathbf{F} = \mathbf{K}^{-1}$.

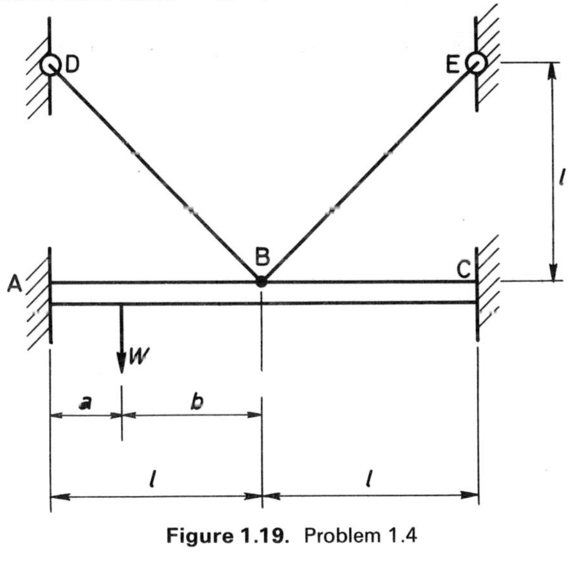

Figure 1.19. Problem 1.4

1.4. A bridge girder ABC, Fig. 1.19, is of uniform flexural rigidity EI, and is encastré at its ends A and C. Additional vertical support is provided by the cables DB and EB which are just taut in the unloaded condition and have an extensional rigidity $AE = 12\sqrt{2}\, EI/l^2$. Using the stiffness method, calculate the cable tensions and the bending moments in the girder at A, B, and C, when the structure carries a load W as shown.

1.5. Construct a stiffness matrix for the frame shown in Fig. 1.20 and hence obtain the rotations of the joints C and D and the bending moments at A and B due to the

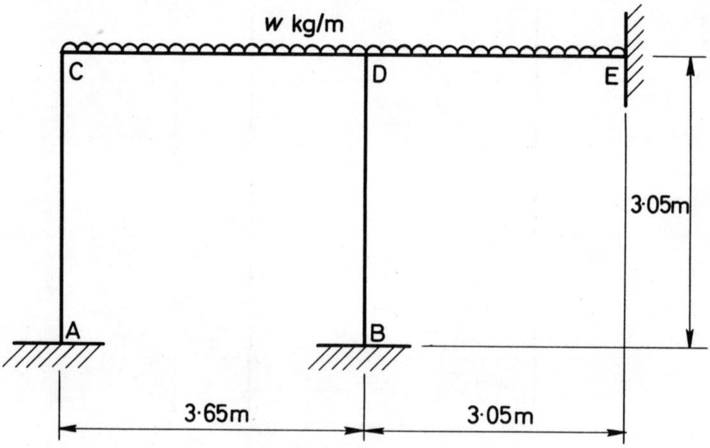

Figure 1.20. Problem 1.5

uniformly distributed load w on the beams.

$$EI\,(\text{beams}) = 3 \times EI\,(\text{columns})$$

1.6. Forces P_x, P_y and moment M_{xy} exist at a point coinciding with the origin of rectangular axes X and Y. The corresponding displacements are x, y, and θ. If the X, Y axes are rotated through angle α, and the transformed forces are $P_x{}^1$, $P_y{}^1$, $M_{xy}{}^1$ with displacements x^1, y^1, θ^1, show that the transformation matrix $\mathbf{T}$ relating forces in the two systems is the same as that relating the displacements and that $\mathbf{T}^T = \mathbf{T}^{-1}$.

2. The Flexibility Method

In chapter 1 we saw how the two alternative approaches of flexibility and stiffness could be used in solving problems in structural analysis. The flexibility method required calculation of flexibility influence coefficients, involving an integration process, and the stiffness method required a calculation of stiffness influence coefficients. These influence coefficients were then used in building up the compatibility or equilibrium equations which had to be solved in the structural analysis. The methods were developed without special attention to the needs of digital computers. The purpose of this chapter is to develop the flexibility method in a suitable form for digital computation. Matrix methods are very suitable for this purpose since digital computers are well equipped to perform the systematic operations of matrix algebra. The flexibility method will be formulated completely in matrix terms as a set of standard operations on the computer. The operations can of course be carried out entirely by hand calculation, but generally they are more suited to digital computers. For hand calculation, other methods are generally more suitable.

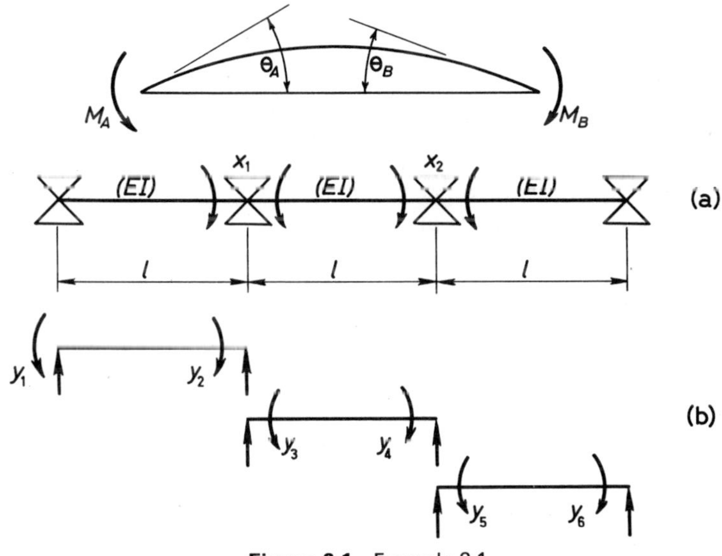

Figure 2.1. Example 2.1

2.1. Flexibility Transformation

Example 2.1. Consider the continuous beam of Fig. 2.1a, in which the internal bending moments x_1 and x_2 at the intermediate supports are chosen as the two redundants. The flexibility matrix of a single span is, from eq. (1.4) with a change of sign convention of M_A,

$$\frac{l}{6EI}\begin{bmatrix} 2 & 1 \\ 1 & 2 \end{bmatrix}$$

If the three spans are first considered to be in an unconnected state, as in Fig. 2.1b, then the flexibility matrix of the system will be

$$\mathbf{F}_y = \frac{l}{6EI}\left[\begin{array}{cc|cc|cc} 2 & 1 & 0 & 0 & 0 & 0 \\ 1 & 2 & 0 & 0 & 0 & 0 \\ \hline 0 & 0 & 2 & 1 & 0 & 0 \\ 0 & 0 & 1 & 2 & 0 & 0 \\ \hline 0 & 0 & 0 & 0 & 2 & 1 \\ 0 & 0 & 0 & 0 & 1 & 2 \end{array}\right] \tag{2.1}$$

If the two sets of forces $\mathbf{X} = \begin{bmatrix} x_1 \\ x_2 \end{bmatrix}$ and $\mathbf{Y} = \begin{bmatrix} y_1 \\ y_2 \\ \vdots \\ y_6 \end{bmatrix}$

are connected by the transformation

$$\mathbf{Y} = \mathbf{T}\mathbf{X} \tag{2.2}$$

then, using the principle of contragredience established in section 1.8, eqns. (1.35) and (1.39), the displacements

$$\mathbf{U}_y = \mathbf{F}_y\mathbf{Y} \tag{2.3}$$

and

$$\mathbf{U}_x = \mathbf{F}_x\mathbf{X} \tag{2.4}$$

are related by

$$\mathbf{U}_x = \mathbf{T}^T\mathbf{U}_y \tag{2.5}$$

Substituting for $\mathbf{Y}$ and $\mathbf{U}_y$ in eq. (2.3)

$$\mathbf{F}_y\mathbf{T}\mathbf{X} = (\mathbf{T}^T)^{-1}\mathbf{U}_x \tag{2.6}$$

$$(\mathbf{T}^T\mathbf{F}_y\mathbf{T})\mathbf{X} = \mathbf{U}_x \tag{2.7}$$

Hence, from eq. (2.4),

$$\mathbf{F}_x = \mathbf{T}^T\mathbf{F}_y\mathbf{T} \tag{2.8}$$

The flexibility of system X has been obtained from the flexibility of system Y by the transformation $\mathbf{T}^T\mathbf{F}_y\mathbf{T}$. This is what is called a congruent transformation, and it follows that if $\mathbf{F}_y$ is a symmetric matrix then the transformed flexibility matrix $\mathbf{F}_x$ is also symmetric. The elements of the transformation matrix $\mathbf{T}$ are, from eq. (2.2), the values of Y corresponding to unit values of X. Thus, in the example of Fig. 2.1,

$$\begin{bmatrix} y_1 \\ y_2 \\ y_3 \\ y_4 \\ y_5 \\ y_6 \end{bmatrix} = \begin{bmatrix} 0 & 0 \\ 1 & 0 \\ 1 & 0 \\ 0 & 1 \\ 0 & 1 \\ 0 & 0 \end{bmatrix} \begin{bmatrix} x_1 \\ x_2 \end{bmatrix} \tag{2.9}$$

Hence,

$$\mathbf{F}_x = \frac{l}{6EI}\left[\begin{array}{cc|cc|cc} 0 & 1 & 1 & 0 & 0 & 0 \\ 0 & 0 & 0 & 1 & 1 & 0 \end{array}\right]\left[\begin{array}{cc|cc|cc} 2 & 1 & 0 & 0 & 0 & 0 \\ 1 & 2 & 0 & 0 & 0 & 0 \\ \hline 0 & 0 & 2 & 1 & 0 & 0 \\ 0 & 0 & 1 & 2 & 0 & 0 \\ \hline 0 & 0 & 0 & 0 & 2 & 1 \\ 0 & 0 & 0 & 0 & 1 & 2 \end{array}\right]\left[\begin{array}{cc} 0 & 0 \\ 1 & 0 \\ \hline 1 & 0 \\ 0 & 1 \\ \hline 0 & 1 \\ 0 & 0 \end{array}\right] \tag{2.10}$$

$$\mathbf{F}_x = \frac{l}{6EI}\begin{bmatrix} 4 & 1 \\ 1 & 4 \end{bmatrix} \tag{2.11}$$

It is left as an exercise for the reader to show by the method of chapter 1 that this is the flexibility matrix of the primary structure, as in Fig. 2.1a.

The flexibility matrix of a given primary system X can thus be obtained from that of the unassembled system Y, using the transformation procedure of eq. (2.8). The elements of the transformation matrix $\mathbf{T}$ are obtained by listing the forces Y in columns corresponding to unit values of the forces X taken in turn, x_1, x_2, etc.

2.2. Compatibility Conditions

The flexibility matrix $\mathbf{F}_x$ represents the displacements in the primary structure at the redundants X due to unit values of the redundants. The displacements at corresponding positions due to the applied loads are represented by the column matrix $\mathbf{U}$. Thus, the compatibility conditions, if there are no discontinuities at the releases in the actual structure, are

$$\mathbf{F}_x\mathbf{X} + \mathbf{U} = \mathbf{0} \tag{2.12}$$

It has been shown how $\mathbf{F}_x$ can be obtained by a matrix transformation process from the individual member flexibilities. It will now be shown that the displacements $\mathbf{U}$ can be obtained by a similar process. Let a vector $\mathbf{R}$ represent a system of external loads producing member forces in the unassembled structure represented by a vector $\mathbf{Y}_0$, and let the relationships between $\mathbf{R}$ and $\mathbf{Y}_0$ be of the form

$$\mathbf{Y}_0 = \mathbf{T}_0\mathbf{R} \tag{2.13}$$

The member displacements $\mathbf{U}_y$ corresponding to member forces $\mathbf{Y}_0$ are

$$\mathbf{U}_y = \mathbf{F}_y\mathbf{Y}_0 \tag{2.14}$$

and from eq. (2.5), the corresponding displacements at the redundants are

$$\mathbf{U}_x = \mathbf{T}^T\mathbf{U}_y \tag{2.15}$$

Substituting for $\mathbf{Y}_0$ and $\mathbf{U}_y$ in eq. (2.14)

$$\mathbf{F}_y\mathbf{T}_0\mathbf{R} = (\mathbf{T}^T)^{-1}\mathbf{U}_x \tag{2.16}$$

Thus

$$(\mathbf{T}^T\mathbf{F}_y\mathbf{T}_0)\mathbf{R} = \mathbf{U}_x \tag{2.17}$$

On the understanding that the displacements at the redundants caused by the applied loads are $\mathbf{U}$, we may omit the subscript in eq. (2.17), and hence

$$\mathbf{U} = (\mathbf{T}^T\mathbf{F}_y\mathbf{T}_0)\mathbf{R} \tag{2.18}$$

Whereas the matrix $\mathbf{T}$ is a *redundant force* transformation matrix, the matrix $\mathbf{T}_0$ is an *external load* transformation matrix. The elements of the matrix $\mathbf{T}_0$ are the member forces, in columns, due to unit values of the applied loads.

Consider the continuous beam of example 2.1, Fig. 2.1a, and assume the applied loading to consist of uniformly distributed loads of intensity w on the end-spans, as in Fig. 2.2a. The primary structure is shown in Fig. 2.2b, and we wish to calculate the displacements at the redundants due to the applied loads. The uniformly distributed loads are equivalent, in effect, to applied moments of $wl^2/8$, as shown in Figs. 2.2f and g. Thus, the applied load vector is

$$\mathbf{R} = \begin{bmatrix} R_1 \\ R_2 \end{bmatrix} = \begin{bmatrix} -wl^2/8 \\ +wl^2/8 \end{bmatrix} \tag{2.19}$$

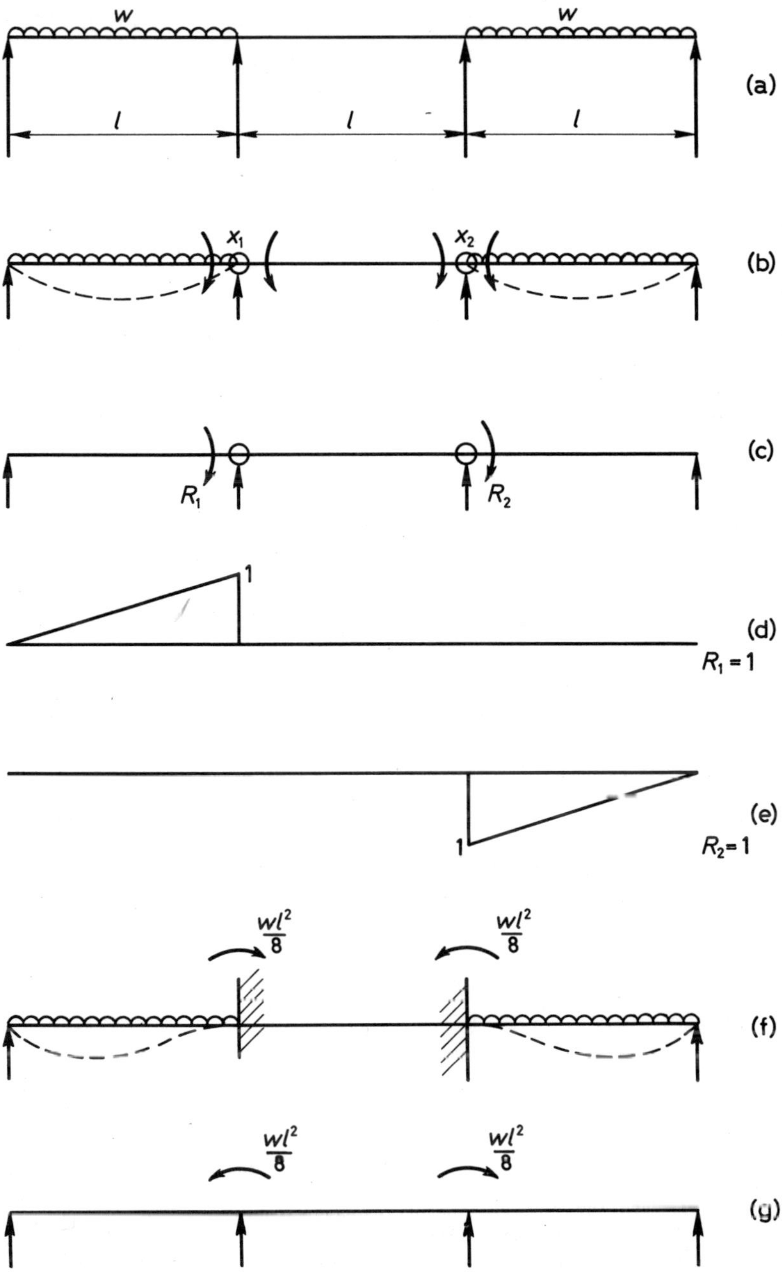
w
w
l
l
l
(a)
X_1
X_2
(b)
R_1
R_2
(c)
1
(d)
$R_1 = 1$
1
(e)
$R_2 = 1$
$\frac{wl^2}{8}$
$\frac{wl^2}{8}$
(f)
$\frac{wl^2}{8}$
$\frac{wl^2}{8}$
(g)

Figure 2.2

The sign convention for R_1 and R_2, implied in Fig. 2.2c, should be noted. The transformation matrix $\mathbf{T}_0$ is then obtained from the end moments in the unassembled spans corresponding to unit values of R_1 and R_2, thus

$$\begin{bmatrix} y_1 \\ y_2 \\ y_3 \\ y_4 \\ y_5 \\ y_6 \end{bmatrix} = \begin{bmatrix} 0 & 0 \\ 1 & 0 \\ 0 & 0 \\ 0 & 0 \\ 0 & -1 \\ 0 & 0 \end{bmatrix} \begin{bmatrix} R_1 \\ R_2 \end{bmatrix} \tag{2.20}$$

Performing the operations indicated in eq. (2.18)

$$\mathbf{U} = \frac{l}{6EI} \begin{bmatrix} 0 & 1 & 1 & 0 & 0 & 0 \\ 0 & 0 & 0 & 1 & 1 & 0 \end{bmatrix} \begin{bmatrix} 2 & 1 & 0 & 0 & 0 & 0 \\ 1 & 2 & 0 & 0 & 0 & 0 \\ 0 & 0 & 2 & 1 & 0 & 0 \\ 0 & 0 & 1 & 2 & 0 & 0 \\ 0 & 0 & 0 & 0 & 2 & 1 \\ 0 & 0 & 0 & 0 & 1 & 2 \end{bmatrix} \begin{bmatrix} 0 & 0 \\ 1 & 0 \\ 0 & 0 \\ 0 & 0 \\ 0 & -1 \\ 0 & 0 \end{bmatrix} \begin{bmatrix} -\dfrac{wl^2}{8} \\ +\dfrac{wl^2}{8} \end{bmatrix} \tag{2.21}$$

$$= \frac{wl^3}{48EI} \begin{bmatrix} -2 \\ -2 \end{bmatrix} \tag{2.22}$$

Using the flexibility matrix already found, eq. (2.11),

$$\frac{l}{6EI} \begin{bmatrix} 4 & 1 \\ 1 & 4 \end{bmatrix} \begin{bmatrix} x_1 \\ x_2 \end{bmatrix} + \frac{wl^3}{48EI} \begin{bmatrix} -2 \\ -2 \end{bmatrix} = \begin{bmatrix} 0 \\ 0 \end{bmatrix} \tag{2.23}$$

The redundants are

$$x_1 = x_2 = \frac{wl^2}{20} \tag{2.24}$$

Solution for multiple load systems. In the above example, the eqns. (2.23) have been solved directly for the redundants x_1 and x_2 corresponding to a single set of loads on the structure. If solutions are required corresponding to different loading conditions, we can of course carry out separate calculations for each set of loads; however, if the number of loading conditions is large, for example in influence line calculations, it is more economical to arrange the computation as follows:

The compatibility conditions are, eq. (2.12),

$$\mathbf{F}_x\mathbf{X} + \mathbf{U} = \mathbf{0}$$

hence

$$\mathbf{X} = -\mathbf{F}_x^{-1}\mathbf{U} \tag{2.25}$$

in which **U** is a rectangular matrix with as many columns as there are loading systems to be considered.

The inverse of the matrix $\mathbf{F}_x$ in eq. (2.11) is easily shown to be

$$\frac{6EI}{l}\begin{bmatrix} \frac{4}{15} & -\frac{1}{15} \\ -\frac{1}{15} & \frac{4}{15} \end{bmatrix}$$

Thus, from eqns. (2.25) and (2.18),

$$\mathbf{X} = -\frac{6EI}{l}\begin{bmatrix} \frac{4}{15} & -\frac{1}{15} \\ -\frac{1}{15} & \frac{4}{15} \end{bmatrix}\frac{l}{6EI}\begin{bmatrix} 2 & 0 \\ 0 & -2 \end{bmatrix}\begin{bmatrix} R_1 \\ R_2 \end{bmatrix}$$

$$\mathbf{X} = -\frac{1}{15}\begin{bmatrix} 8 & 2 \\ -2 & -8 \end{bmatrix}\begin{bmatrix} R_1 \\ R_2 \end{bmatrix} \tag{2.26}$$

It will be noted that this process involves a matrix inversion and a matrix multiplication. This will often be more economical than a process involving the solution of separate sets of equations corresponding to the different loading conditions. In this way, advantage is taken of the fact that the flexibility matrix $\mathbf{F}_x$ is independent of the loading on the structure and depends only on the properties of the structure and the release system adopted.

2.3. Calculation of Member Forces

Once the redundants have been calculated corresponding to any particular set of applied loads, the next stage in the calculation will normally be to obtain the state of stress in the members. This is now a problem in statics since the statically indeterminate part of the problem has been disposed of. The matrix formulation of the flexibility method will now be extended to include the calculation of member forces.

Equation (2.2) can be used to obtain the member forces **TX** caused by the redundants **X**.

The member forces caused by the applied loads acting on the primary structure, the *particular* solutions, are [eq. (2.13)] $\mathbf{T}_0\mathbf{R}$. Thus, the total member forces are

$$\mathbf{S} = \mathbf{T}_0\mathbf{R} + \mathbf{TX} \tag{2.27}$$

Equation (2.27) represents the total solution for the end bending moments in the members for the structure carrying external loads specified in a loading vector **R**. If these loads are the statical equivalent of a set of internal and fixed-end bending moments in a certain member, then this system of moments must be added to that obtained by eq. (2.27) to obtain the final distribution of bending moments in the structure.

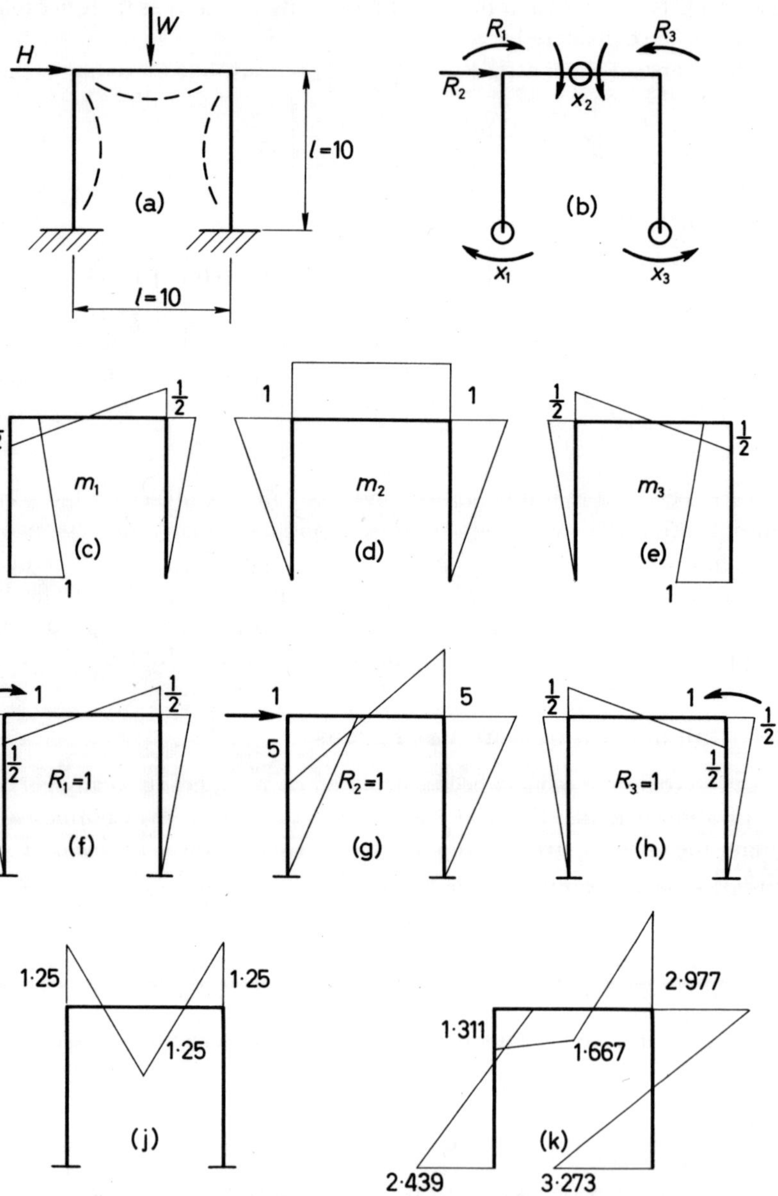

Figure 2.3. Example 2.2

Example 2.2. The structure shown in Fig. 2.3 has three redundancies taken to be x_1, x_2, x_3, Fig. 2.3b. The bending moment diagrams due to unit values of the redundancies and the loads $R_{1,2,3}$, are shown in c, d, e, and f, g, h.

The flexibility matrix of the unassembled members is

$$\mathbf{F}_y = \frac{c}{6}\begin{bmatrix} 2 & 1 & 0 & 0 & 0 & 0 \\ 1 & 2 & 0 & 0 & 0 & 0 \\ 0 & 0 & 2 & 1 & 0 & 0 \\ 0 & 0 & 1 & 2 & 0 & 0 \\ 0 & 0 & 0 & 0 & 2 & 1 \\ 0 & 0 & 0 & 0 & 1 & 2 \end{bmatrix} \tag{2.28}$$

where $c = l/EI$. The sign conventions for bending moments in the three members are that bending is positive if the curvature is as shown in diagram (a).

The transformation matrices required are

$$\mathbf{T} = \begin{bmatrix} 1 & 0 & 0 \\ \frac{1}{2} & -1 & -\frac{1}{2} \\ \frac{1}{2} & -1 & -\frac{1}{2} \\ -\frac{1}{2} & -1 & \frac{1}{2} \\ -\frac{1}{2} & -1 & \frac{1}{2} \\ 0 & 0 & 1 \end{bmatrix}; \quad \mathbf{T}_0 = \begin{bmatrix} 0 & 0 & 0 \\ -\frac{1}{2} & 5 & -\frac{1}{2} \\ \frac{1}{2} & 5 & -\frac{1}{2} \\ -\frac{1}{2} & -5 & \frac{1}{2} \\ -\frac{1}{2} & -5 & -\frac{1}{2} \\ 0 & 0 & 0 \end{bmatrix} \tag{2.29}$$

Hence

$$\mathbf{F}_x = \mathbf{T}^T\mathbf{F}_y\mathbf{T} = \frac{c}{6}\begin{bmatrix} 4\cdot5 & -1 & -2\cdot5 \\ -1 & 10 & -1 \\ -2\cdot5 & -1 & 4\cdot5 \end{bmatrix} \tag{2.30}$$

and

$$\mathbf{U} = (\mathbf{T}^T\mathbf{F}_y\mathbf{T}_0)\mathbf{R} = \frac{c}{6}\begin{bmatrix} 0 & 20 & -1 \\ 2 & 0 & 2 \\ -1 & -20 & 0 \end{bmatrix}\begin{bmatrix} R_1 \\ R_2 \\ R_3 \end{bmatrix} \tag{2.31}$$

$$\mathbf{F}_x^{-1} = \frac{6}{c}\begin{bmatrix} 0\cdot3493 & 0\cdot0556 & 0\cdot2065 \\ 0\cdot0556 & 0\cdot1113 & 0\cdot0556 \\ 0\cdot2065 & 0\cdot0556 & 0\cdot3493 \end{bmatrix} \tag{2.32}$$

Hence $$\mathbf{X} = -\mathbf{F}_x^{-1}\mathbf{U}$$

$$= -\begin{bmatrix} -0{\cdot}0953 & 2{\cdot}856 & -0{\cdot}2381 \\ 0{\cdot}1670 & 0 & 0{\cdot}1670 \\ -0{\cdot}2381 & -2{\cdot}856 & -0{\cdot}0953 \end{bmatrix} \begin{bmatrix} R_1 \\ R_2 \\ R_3 \end{bmatrix} \tag{2.33}$$

and $$R_1 = \frac{Wl}{8}; \qquad R_2 = H; \qquad R_3 = \frac{Wl}{8} \tag{2.34}$$

hence $$\mathbf{X} = \begin{bmatrix} +0{\cdot}04167Wl - 2{\cdot}856H \\ -0{\cdot}04175Wl \\ +0{\cdot}04167Wl + 2{\cdot}856H \end{bmatrix} \tag{2.35}$$

If we now allot unit value to W and H

$$\mathbf{X} = \begin{bmatrix} -2{\cdot}439 \\ -0{\cdot}4175 \\ +3{\cdot}273 \end{bmatrix} \tag{2.36}$$

and the member forces are

$$\mathbf{S} = \mathbf{T}_0\mathbf{R} + \mathbf{TX}$$

$$= \begin{bmatrix} 0 & 0 & 0 \\ -\frac{1}{2} & 5 & -\frac{1}{2} \\ \frac{1}{2} & 5 & -\frac{1}{2} \\ -\frac{1}{2} & -5 & \frac{1}{2} \\ -\frac{1}{2} & -5 & -\frac{1}{2} \\ 0 & 0 & 0 \end{bmatrix} \begin{bmatrix} \frac{10}{8} \\ 1 \\ \frac{10}{8} \end{bmatrix} + \begin{bmatrix} 1 & 0 & 0 \\ \frac{1}{2} & -1 & -\frac{1}{2} \\ \frac{1}{2} & -1 & -\frac{1}{2} \\ -\frac{1}{2} & -1 & \frac{1}{2} \\ -\frac{1}{2} & -1 & \frac{1}{2} \\ 0 & 0 & 1 \end{bmatrix} \begin{bmatrix} -2{\cdot}439 \\ -0{\cdot}4175 \\ +3{\cdot}273 \end{bmatrix}$$

$$= \begin{bmatrix} 0 \\ 3{\cdot}75 \\ 5{\cdot}00 \\ -5{\cdot}00 \\ -6{\cdot}25 \\ 0 \end{bmatrix} + \begin{bmatrix} -2{\cdot}439 \\ -2{\cdot}439 \\ -2{\cdot}439 \\ +3{\cdot}273 \\ +3{\cdot}273 \\ +3{\cdot}273 \end{bmatrix} = \begin{bmatrix} -2{\cdot}439 \\ +1{\cdot}311 \\ +2{\cdot}561 \\ -1{\cdot}727 \\ -2{\cdot}977 \\ +3{\cdot}273 \end{bmatrix} \tag{2.37}$$

This constitutes the solution for loading system $R_{1,2,3}$, as in Fig. 2.3b. To this has to be added the system of internal bending moments due to the constraining (fixed-end) moments and the applied mid-span load, as in Fig. 2.3j. The final bending moment diagram is then as shown in Fig. 2.3k.

2.4. Support Displacements, Prestrain, and Thermal Effects

In the basic compatibility equations $\mathbf{FX} + \mathbf{U} = \mathbf{0}$, the matrix $\mathbf{U}$ represents the displacements in the primary structure at the positions of the releases caused by the applied loads. Such displacements can, however, be caused by other effects, such as change of temperature of some part of the structure or a settlement of one of the supports.

A more general form for the compatibility conditions is, therefore,

$$\mathbf{F}_x\mathbf{X} + \mathbf{U}_R + \mathbf{U}_T + \mathbf{U}_S = \Delta \tag{2.38}$$

where $\mathbf{U}_R$, $\mathbf{U}_T$, and $\mathbf{U}_S$ are the discontinuities produced at the releases of the primary structure due to loading, temperature, and settlement or prestrain respectively. The vector Δ on the right-hand side of the equation represents the final discontinuity effect at the releases. The elements of Δ will be positive when the final discontinuity is in the same direction as the positive redundant.

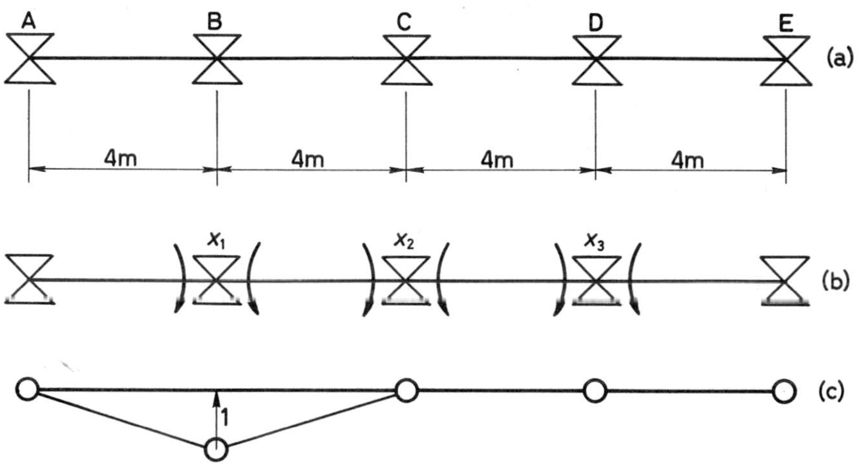

Figure 2.4. Example 2.3

Example 2.3. The four span continuous beam of Fig. 2.4a is composed of a uniform section member having $EI = 133 \times 10^9$ kN mm^2. Find the bending moment distribution resulting from a unit vertical settlement of support B.

Adopting the release system shown in Fig. 2.4b, the compatibility conditions are $\mathbf{F}_x\mathbf{X} + \mathbf{U}_S = \mathbf{0}$, and the elements in $\mathbf{U}_S$ are the rotations in the primary structure at the position of the releases. $\Delta = 0$ here since there is no discontinuity at the releases in the final state of the structure.

The unit vertical displacement at B will produce a rotation at B of $1/l_{AB} + 1/l_{BC} = \frac{1}{2}$. The rotation is positive in this case since one span rotates relative to the other, in a sense corresponding to the positive direction of the redundant x_1. The rotation at C is $-1/l_{BC} = -\frac{1}{4}$, negative since the rotation is counter to the positive direction of x_2. For the chosen primary structure, the rotation at D is unaffected by the settlement of B.

Thus,

$$\mathbf{U}_S = \begin{bmatrix} +\frac{1}{2} \\ -\frac{1}{4} \\ 0 \end{bmatrix} \tag{2.39}$$

The flexibility matrix of the primary structure is

$$\mathbf{F}_x = \frac{1}{133 \times 10^3} \begin{bmatrix} \frac{8}{3} & \frac{2}{3} & 0 \\ \frac{2}{3} & \frac{8}{3} & \frac{2}{3} \\ 0 & \frac{2}{3} & \frac{8}{3} \end{bmatrix} \tag{2.40}$$

Thus

$$\mathbf{X} = -\mathbf{F}_x^{-1}\mathbf{U}_S = 3 \times 133 \times 10^3 \begin{bmatrix} 0{\cdot}133 & -0{\cdot}0357 & 0{\cdot}0089 \\ -0{\cdot}0357 & 0{\cdot}143 & -0{\cdot}0357 \\ 0{\cdot}0089 & -0{\cdot}0357 & 0{\cdot}133 \end{bmatrix} \begin{bmatrix} -\frac{1}{2} \\ +\frac{1}{4} \\ 0 \end{bmatrix}$$

$$= \begin{bmatrix} -30\,095 \\ +21\,376 \\ -\ 5\,337 \end{bmatrix} \text{(kNm/m units)} \tag{2.41}$$

The process of converting the support yield into rotations at the releases can be carried out using a transformation matrix. If we have two release systems X_1 and X_2, Fig. 2.5, with compatibility conditions $\mathbf{F}_1\mathbf{X}_1 + \mathbf{U}_S = \mathbf{0}$

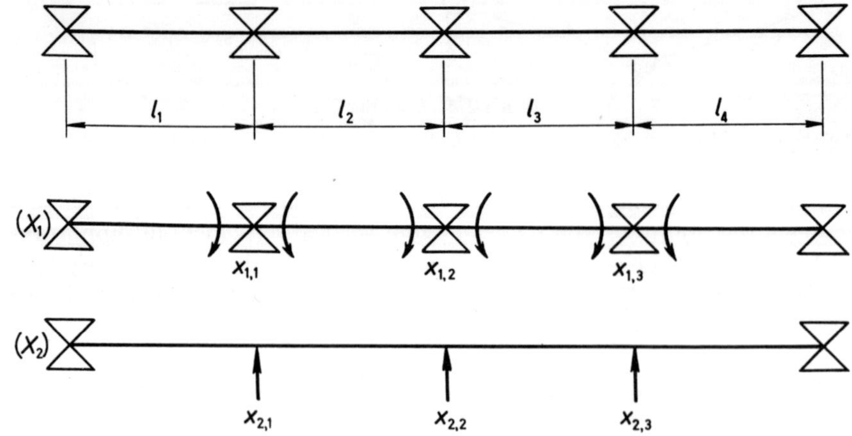

Figure 2.5. Alternative primary structures

and $\mathbf{F}_2\mathbf{X}_2 = \Delta$, and if $\mathbf{X}_2 = \mathbf{T}\mathbf{X}_1$, then by the principle of contragredience, section 1.8,

$$-\mathbf{U}_S = \mathbf{T}^T \Delta \tag{2.42}$$

Thus

$$\mathbf{F}_1\mathbf{X}_1 = \mathbf{T}^T\Delta = -\mathbf{U}_S \tag{2.43}$$

The transformation matrix **T** relates the two systems of redundants $\mathbf{X}_1$ and $\mathbf{X}_2$, that is,

$$\begin{bmatrix} x_{2,1} \\ x_{2,2} \\ x_{2,3} \end{bmatrix} = \begin{bmatrix} (1/l_1 + 1/l_2) & -1/l_2 & 0 \\ -1/l_2 & (1/l_2 + 1/l_3) & -1/l_3 \\ 0 & -1/l_3 & (1/l_3 + 1/l_4) \end{bmatrix} \begin{bmatrix} x_{1,1} \\ x_{1,2} \\ x_{1,3} \end{bmatrix} \tag{2.44}$$

and

$$\mathbf{T}^T \begin{bmatrix} -1 \\ 0 \\ 0 \end{bmatrix} = \begin{bmatrix} -\frac{1}{2} \\ \frac{1}{4} \\ 0 \end{bmatrix} = -\mathbf{U}_S \tag{2.45}$$

which agrees with eq. (2.39).

The treatment of signs needs special care. The elements in the matrices in eq. (2.38) are all displacements, or discontinuities, at the positions of the redundants. The sign convention throughout is that a displacement is positive if in the same direction as the corresponding redundant. The right-hand side matrix Δ is the resultant discontinuity, frequently zero, when all the effects have been added. The elements of Δ are positive if in the positive directions of X.

2.5. Choice of Primary Structure

Having determined the degree of statical indeterminacy of a structure to be analysed by the flexibility method, a number of considerations should be borne in mind when choosing a system of releases.

(a) The releases should be distributed throughout the structure in such a way that the complete primary structure is statically determinate. If part of the primary structure remains indeterminate it follows that some other part has become a mechanism, assuming the correct number of releases to have been introduced.

(b) Since a substantial amount of statical analysis may have to be carried out on the primary structure this should, if possible, be chosen so that the statical analysis is as simple as possible.

(c) The accuracy with which the compatibility equations $\mathbf{FX} + \mathbf{U} = \mathbf{0}$ can be solved depends to a large extent on the '*conditioning*' of the matrix **F**. For well conditioned equations, the diagonal elements should be large in comparison with the off-diagonal elements. This is seen if we examine a set of two equations:

$$\begin{bmatrix} a_1 & b \\ b & a_2 \end{bmatrix} \begin{bmatrix} x_1 \\ x_2 \end{bmatrix} = \begin{bmatrix} c_1 \\ c_2 \end{bmatrix} \tag{2.46}$$

The solution for x_1 is

$$x_1 = \frac{c_1 - (b/a_2)c_2}{a_1 - b^2/a_2} \tag{2.47}$$

If, in the denominator of eq. (2.47), b^2/a_2 approximates to a_1, then the accuracy of evaluation of x_1 will be reduced unless more significant figures are used. It can happen that a set of equations is so ill-conditioned that it is impracticable to use the number of significant digits required to obtain an acceptable accuracy. Generally, with structural problems it is desirable and possible to avoid ill-conditioned equations. If, in eq. (2.47), b is significantly smaller than a_1 and a_2, no difficulty arises. For the diagonal elements to be dominant in the matrix the release system should be chosen in such a way that any redundant produces a larger displacement at its own release than at any other release. In system (a) of Fig. 2.6, the redundant x_1 will produce a larger rotation at x_1 than at x_2. Furthermore, the limited overlap of the m diagrams ensures $f_{13}, f_{14} = 0$. In system (b) it is clear that redundant y_1 will produce a larger displacement at y_2 than it does at y_1. It follows that system (a) will give well-conditioned equations and system (b) ill-conditioned equations.

The above consideration may be applied to give a qualitative assessment of the conditioning of a set of equations, a quantitative assessment may be obtained as follows:

The set of equations is first *normalized* by dividing the coefficients in each equation by the square root of the sum of the squares of the coefficients in that equation. The determinant of the resulting *normalized* matrix is then evaluated. If this process is carried out on the 'ideal' set of equations,

$$f_{11}x_1 = b_1$$
$$f_{22}x_2 = b_2$$

then the *normalized* version of matrix $\mathbf{F}$ is

$$\mathbf{F} = \begin{bmatrix} 1 & 0 \\ 0 & 1 \end{bmatrix}$$

and the determinant of $\mathbf{F} = |\mathbf{F}| = 1$.

When the process is applied to a given flexibility matrix the value of the resulting determinant is thus compared with unity. Relatively small values of $|\mathbf{F}|$ indicate that the system is ill conditioned. For the continuous beam of Fig. 2.6, the two release systems (a) and (b) give values of $|\mathbf{F}|$ equal to 0·685 and 0·0001 respectively, indicating that (a) is well conditioned and (b) ill conditioned.

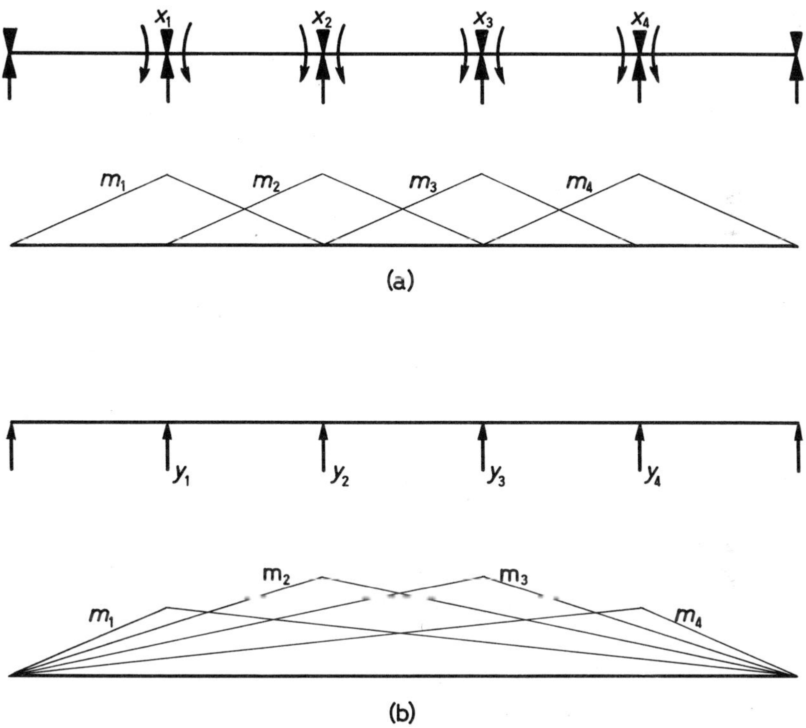

Figure 2.6. Alternative primary structures

In continuous beams and rigid frames moment releases are preferable to separations.

2.6. Calculation of Deflections

The matrix formulation of the flexibility method will now be extended to include the calculation of structural deflections. The method will provide for the calculation of deflections at positions corresponding to those of the applied load vector **R**. If a deflection is required for which there is no corresponding applied load this can be accommodated simply by enlarging the vector **R** by a zero element in the appropriate position. An illustration of this type of calculation will be provided in example 2.4.

It will be recalled that the deflections in the primary structure at the positions of the releases due to unit redundants were obtained as, eq. (2.8),

$$\mathbf{F}_x = \mathbf{T}^T\mathbf{F}_y\mathbf{T}$$

and that $\mathbf{F}_x$ is the primary structure flexibility matrix. A similar matrix treatment can be used to determine $\mathbf{F}$, the flexibility matrix of the actual structure.

If the equation for the member forces, eq. (2.27), is re-written

$$\begin{aligned}\mathbf{S} &= \mathbf{T}_0\mathbf{R} + \mathbf{T}\mathbf{X}_1\mathbf{R} \\ &= (\mathbf{T}_0 + \mathbf{T}\mathbf{X}_1)\mathbf{R} = \mathbf{B}\mathbf{R} \qquad (2.48)\end{aligned}$$

then the matrix $\mathbf{X}_1$ contains the values of the redundants corresponding to unit applied loads and the matrix $\mathbf{B}$ contains the member forces due to unit applied loads and the corresponding redundants.

$$\mathbf{B} = (\mathbf{T}_0 + \mathbf{T}\mathbf{X}_1) \qquad (2.49)$$

The relationship between member forces and applied loads for the primary structure is

$$\mathbf{Y}_0 = \mathbf{T}_0\mathbf{R} \qquad (2.50)$$

The corresponding member displacements are

$$\mathbf{U}_y = \mathbf{F}_y\mathbf{S} \qquad (2.51)$$

If the structure displacements due to loads $\mathbf{R}$ are

$$\mathbf{r} = \mathbf{F}\mathbf{R} \qquad (2.52)$$

then from the principle of contragredience, section 1.8,

$$\mathbf{r} = \mathbf{T}_0{}^T\mathbf{U}_y \qquad (2.53)$$

Thus, substituting in eq. (2.52) for $\mathbf{r}$ and $\mathbf{U}_y$,

$$\mathbf{T}_0{}^T\mathbf{F}_y\mathbf{B}\mathbf{R} = \mathbf{F}\mathbf{R} \qquad (2.54)$$

Thus, the flexibility matrix of the structure associated with the positions and directions implied in $\mathbf{R}$ is

$$\mathbf{F} = \mathbf{T}_0{}^T\mathbf{F}_y\mathbf{B} \qquad (2.55)$$

and the displacements due to $\mathbf{R}$ are

$$\mathbf{r} = \mathbf{F}\mathbf{R} \qquad (2.56)$$

Example 2.4. Example 2.2 will now be continued to the stage of calculating deflections

$\mathbf{B} = \mathbf{T}_0 + \mathbf{T}\mathbf{X}_1$

$$= \begin{bmatrix} 0 & 0 & 0 \\ -\frac{1}{2} & 5 & -\frac{1}{2} \\ \frac{1}{2} & 5 & -\frac{1}{2} \\ -\frac{1}{2} & -5 & \frac{1}{2} \\ -\frac{1}{2} & -5 & -\frac{1}{2} \\ 0 & 0 & 0 \end{bmatrix} + \begin{bmatrix} 1 & 0 & 0 \\ \frac{1}{2} & -1 & -\frac{1}{2} \\ \frac{1}{2} & -1 & -\frac{1}{2} \\ -\frac{1}{2} & -1 & \frac{1}{2} \\ -\frac{1}{2} & -1 & \frac{1}{2} \\ 0 & 0 & 1 \end{bmatrix} \begin{bmatrix} +0{\cdot}0953 & -2{\cdot}856 & +0{\cdot}2381 \\ -0{\cdot}1670 & 0 & -0{\cdot}1670 \\ +0{\cdot}2381 & +2{\cdot}856 & +0{\cdot}0953 \end{bmatrix} \tag{2.57}$$

$$= \begin{bmatrix} 0{\cdot}0953 & -2{\cdot}856 & 0{\cdot}2381 \\ -0{\cdot}4045 & 2{\cdot}144 & -0{\cdot}2615 \\ 0{\cdot}5955 & 2{\cdot}146 & -0{\cdot}2615 \\ -0{\cdot}2615 & -2{\cdot}146 & 0{\cdot}5955 \\ -0{\cdot}2615 & -2{\cdot}146 & -0{\cdot}4045 \\ 0{\cdot}2381 & 2{\cdot}856 & 0{\cdot}0955 \end{bmatrix}$$

$$\mathbf{F} = \mathbf{T}_0{}^T\mathbf{F}_y\mathbf{B} = \frac{10}{6EI}\begin{bmatrix} 0{\cdot}929 & 2{\cdot}14 & 0{\cdot}071 \\ 2{\cdot}14 & 35{\cdot}78 & -2{\cdot}14 \\ 0{\cdot}071 & -2{\cdot}14 & 0{\cdot}929 \end{bmatrix} \tag{2.58}$$

$$\mathbf{r} = \mathbf{F}\mathbf{R} = \frac{1}{EI}\begin{bmatrix} 5{\cdot}65 \\ 59{\cdot}6 \\ -1{\cdot}49 \end{bmatrix} \tag{2.59}$$

As an illustration of the method in the calculation of a displacement at an unloaded point in a structure, consider the frame of Fig. 2.3a, with the vertical load W removed. The displacements in the directions of $R_{1,2,3}$ may be calculated corresponding to the load H alone, by setting

$$\mathbf{R} = \begin{bmatrix} 0 \\ 1 \\ 0 \end{bmatrix}$$

The flexibility matrix **F**, pertaining to the directions $R_{1,2,3}$, is independent of the actual values of the applied loads, and is still given by eq. (2.58). Thus

$$\mathbf{r} = \mathbf{FR}$$

$$= \frac{1}{EI}\begin{bmatrix} 3{\cdot}57 \\ 59{\cdot}6 \\ -3{\cdot}57 \end{bmatrix}$$

2.7. Plane Truss Analysis

The development of the flexibility method has been illustrated with examples of structures which carry loads principally by the resistance of their members to bending. The method is equally suitable for trusses whose load-carrying capacity results from the axial loads carried by the members. The member flexibilities now arise from axial deformations.

Statically determinate trusses. Here the relationships between member forces and applied loads are obtainable purely from considerations of static equilibrium, and are expressed by eq. (2.13)

$$\mathbf{Y}_0 = \mathbf{T}_0\mathbf{R}$$

in which $\mathbf{Y}_0$ is a matrix of member forces, **R** is a matrix of applied loads, and $\mathbf{T}_0$ a force transformation matrix. The applied loads and displacements are related by

$$\mathbf{r} = (\mathbf{T}_0{}^T\mathbf{F}_y\mathbf{T}_0)\mathbf{R} \tag{2.60}$$

in which $\mathbf{T}_0{}^T\mathbf{F}_y\mathbf{T}_0$ is the flexibility matrix of the primary and actual structures, since in this case there is no difference.

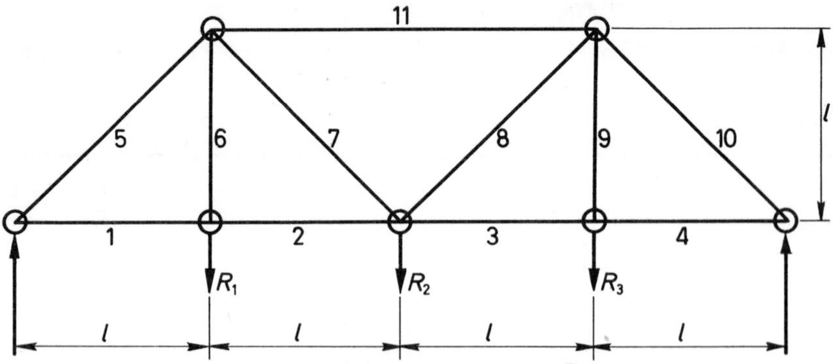

Figure 2.7. Example 2.5

Example 2.5, Fig. 2.7. AE = constant for all members. Calculate vertical deflections of load points. Tensile forces y are considered positive.

$$\mathbf{Y}_0 = \mathbf{T}_0\mathbf{R}, \text{ i.e., } \begin{bmatrix} y_1 \\ y_2 \\ y_3 \\ y_4 \\ y_5 \\ y_6 \\ y_7 \\ y_8 \\ y_9 \\ y_{10} \\ y_{11} \end{bmatrix} = \begin{bmatrix} \frac{3}{4} & \frac{1}{2} & \frac{1}{4} \\ \frac{3}{4} & \frac{1}{2} & \frac{1}{4} \\ \frac{1}{4} & \frac{1}{2} & \frac{3}{4} \\ \frac{1}{4} & \frac{1}{2} & \frac{3}{4} \\ -3/2\sqrt{2} & -1/\sqrt{2} & -1/2\sqrt{2} \\ 1 & 0 & 0 \\ -1/2\sqrt{2} & 1/\sqrt{2} & 1/2\sqrt{2} \\ 1/2\sqrt{2} & 1/\sqrt{2} & -1/2\sqrt{2} \\ 0 & 0 & 1 \\ -1/2\sqrt{2} & -1/\sqrt{2} & -3/2\sqrt{2} \\ -\frac{1}{2} & -1 & -\frac{1}{2} \end{bmatrix} \begin{bmatrix} R_1 \\ R_2 \\ R_3 \end{bmatrix}$$

$$\mathbf{F}_y\mathbf{Y}_0 = \mathbf{U}_y = \frac{l}{AE} \left\lceil \begin{matrix} 1 \\ 1 \\ 1 \\ 1 \\ \sqrt{2} \\ 1 \\ \sqrt{2} \\ \sqrt{2} \\ 1 \\ \sqrt{2} \\ 2 \end{matrix} \right\rfloor \begin{bmatrix} y_1 \\ y_2 \\ y_3 \\ y_4 \\ y_5 \\ y_6 \\ y_7 \\ y_8 \\ y_9 \\ y_{10} \\ y_{11} \end{bmatrix}$$

(The brackets ⌈ ⌋ indicate a diagonal matrix.)

$\mathbf{F}_R = \mathbf{T}_0{}^T\mathbf{F}_y\mathbf{T}_0$ = flexibility matrix pertaining to the directions of $R_{1,2,3}$

$$= \mathbf{T}_0{}^T \frac{l}{AE}\begin{bmatrix} \frac{3}{4} & \frac{1}{2} & \frac{1}{4} \\ \frac{3}{4} & \frac{1}{2} & \frac{1}{4} \\ \frac{1}{4} & \frac{1}{2} & \frac{3}{4} \\ \frac{1}{4} & \frac{1}{2} & \frac{3}{4} \\ -\frac{3}{2} & -1 & -\frac{1}{2} \\ 1 & 0 & 0 \\ -\frac{1}{2} & 1 & \frac{1}{2} \\ \frac{1}{2} & \frac{1}{2} & -\frac{1}{2} \\ 0 & 0 & 1 \\ -\frac{1}{2} & -1 & -\frac{3}{2} \\ -1 & -2 & -1 \end{bmatrix} = \frac{l}{4AE}\begin{bmatrix} (11+6\sqrt{2}) & 2(4+2\sqrt{2}) & (5+2\sqrt{2}) \\ 2(4+2\sqrt{2}) & 4(3+2\sqrt{2}) & 2(4+2\sqrt{2}) \\ (5+2\sqrt{2}) & 2(4+2\sqrt{2}) & (11+6\sqrt{2}) \end{bmatrix}$$

Thus, the load point deflections are

$$\mathbf{U}_R = \mathbf{F}_R\mathbf{R}$$

$$= \frac{l}{4AE}\begin{bmatrix} R_1(11+6\sqrt{2}) + 2R_2(4+2\sqrt{2}) + R_3(5+2\sqrt{2}) \\ 2R_1(4+2\sqrt{2}) + 4R_2(3+2\sqrt{2}) + 2R_3(4+2\sqrt{2}) \\ R_1(5+2\sqrt{2}) + 2R_2(4+2\sqrt{2}) + R_3(11+6\sqrt{2}) \end{bmatrix}$$

Statically indeterminate trusses. The statical indeterminacy n_S is determined, and a system of redundants $x_1 \ldots x_{n_s}$ selected. These are conveniently the forces in certain of the members of the truss and so the primary structure is one in which these members are omitted. The member forces **Y** and redundants **X** and loads **R** are related by eqns. (2.2), (2.13)

$$\mathbf{Y} = \mathbf{TX}, \qquad \mathbf{Y}_0 = \mathbf{T}_0\mathbf{R}$$

and the flexibility of the primary structure corresponding to **X** is calculated, eq. (2.8),

$$\mathbf{F}_x = \mathbf{T}^T\mathbf{F}_y\mathbf{T}$$

The primary structure displacements at the positions of the redundants due to the applied loads are, eq. (2.18),

$$\mathbf{U}_x = \mathbf{U} = (\mathbf{T}^T\mathbf{F}_y\mathbf{T}_0)\mathbf{R}$$

The compatibility conditions are

$$\mathbf{F}_x\mathbf{X} + \mathbf{U} = \mathbf{0}$$

hence

$$\mathbf{X} = -\mathbf{F}_x{}^{-1}\mathbf{U}$$

Example 2.6, Fig. 2.8. AE = constant for all members = 1×10^6 kN. Diagonal x_2 is $l/10{,}000$ short.

$$\mathbf{Y}_0 = \mathbf{T}_0\mathbf{R} \qquad\qquad \mathbf{Y} = \mathbf{TX}$$

$$\begin{bmatrix} y_1 \\ y_2 \\ y_3 \\ y_4 \\ y_5 \\ y_6 \\ y_7 \\ y_8 \\ y_9 \\ y_{10} \\ y_{11} \end{bmatrix} = \begin{bmatrix} 0 \\ 0 \\ -\frac{1}{2} \\ 0 \\ 1/\sqrt{2} \\ 0 \\ 1/\sqrt{2} \\ 0 \\ -\frac{1}{2} \\ -\frac{1}{2} \\ -\frac{1}{2} \end{bmatrix} [\mathbf{R}]; \quad \begin{bmatrix} y_1 \\ y_2 \\ y_3 \\ y_4 \\ y_5 \\ y_6 \\ y_7 \\ y_8 \\ y_9 \\ y_{10} \\ y_{11} \end{bmatrix} = \begin{bmatrix} -1/\sqrt{2} & 0 \\ 0 & -1/\sqrt{2} \\ -1/\sqrt{2} & 0 \\ 1 & 0 \\ 1 & 0 \\ -1/\sqrt{2} & -1/\sqrt{2} \\ 0 & 1 \\ 0 & 1 \\ 0 & -1/\sqrt{2} \\ -1/\sqrt{2} & 0 \\ 0 & -1/\sqrt{2} \end{bmatrix} \begin{bmatrix} x_1 \\ x_2 \end{bmatrix}$$

$$\mathbf{F}_y = \frac{l}{AE} \begin{bmatrix} 1 & & & & & & & & & & \\ & 1 & & & & & & & & & \\ & & 1 & & & & & & & & \\ & & & \sqrt{2} & & & & & & & \\ & & & & \sqrt{2} & & & & & & \\ & & & & & 1 & & & & & \\ & & & & & & \sqrt{2} & & & & \\ & & & & & & & \sqrt{2} & & & \\ & & & & & & & & 1 & & \\ & & & & & & & & & 1 & \\ & & & & & & & & & & 1 \end{bmatrix}$$

$$\mathbf{U}_R = (\mathbf{T}^T\mathbf{F}_y\mathbf{T}_0)\mathbf{R} \qquad\qquad \mathbf{F}_x = \mathbf{T}^T\mathbf{F}_y\mathbf{T}$$

$$= \frac{lR}{AE}\begin{bmatrix} 1 + 1/\sqrt{2} \\ 1 + 1/\sqrt{2} \end{bmatrix}; \qquad \mathbf{F}_x = \frac{l}{AE}\begin{bmatrix} 2 + 2\sqrt{2} & \frac{1}{2} \\ \frac{1}{2} & 2 + 2\sqrt{2} \end{bmatrix}$$

$$\mathbf{F}_x\mathbf{X} + \mathbf{U}_R + \mathbf{U}_S = \mathbf{0}$$

$$\frac{l}{2AE}\begin{bmatrix} 4+4\sqrt{2} & 1 \\ 1 & 4+4\sqrt{2} \end{bmatrix}\begin{bmatrix} x_1 \\ x_2 \end{bmatrix} + \frac{Rl}{2AE}\begin{bmatrix} 2+\sqrt{2} \\ 2+\sqrt{2} \end{bmatrix} - l\begin{bmatrix} 0 \\ 1/10{,}000 \end{bmatrix} = \begin{bmatrix} 0 \\ 0 \end{bmatrix}$$

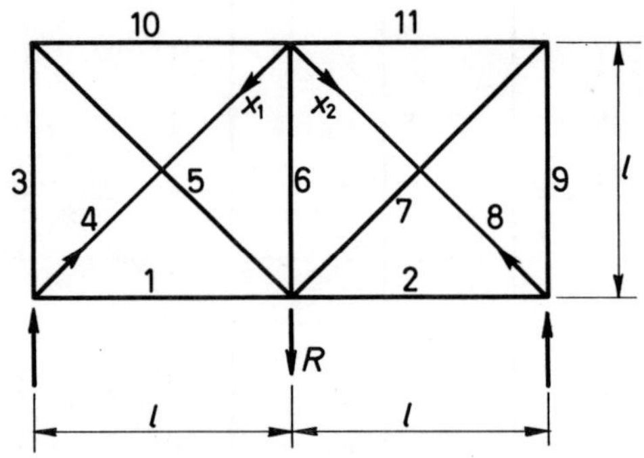

Figure 2.8. Example 2.6

The equations may be solved for the effects of the applied load and prestrain separately; thus, due to the applied load R,

$$x_1 = x_2 = -\left(\frac{2+\sqrt{2}}{5+4\sqrt{2}}\right)R$$

and due to the prestrain,

$$x_1 = \left(\frac{-2}{47+32\sqrt{2}}\right)10^2 \text{ kN}$$

$$x_2 = \frac{8(1+\sqrt{2})10^2}{(47+32\sqrt{2})}\text{ kN}$$

The member forces may be calculated from eq. (2.48)

$$\mathbf{S} = \mathbf{T}_0\mathbf{R} + \mathbf{TX}$$

PROBLEMS

2.1. When the frame shown in Fig. 2.9 carries the loading shown, the column foot D rotates 0·002 radian clockwise and sinks vertically 18·8 mm. The flexural rigidity of the beam is twice that of the columns. Use the flexibility method to calculate the reactions at D. Take $E = 200$ kN/mm^2 and I (column) $= 333 \times 10^6$ mm^4.

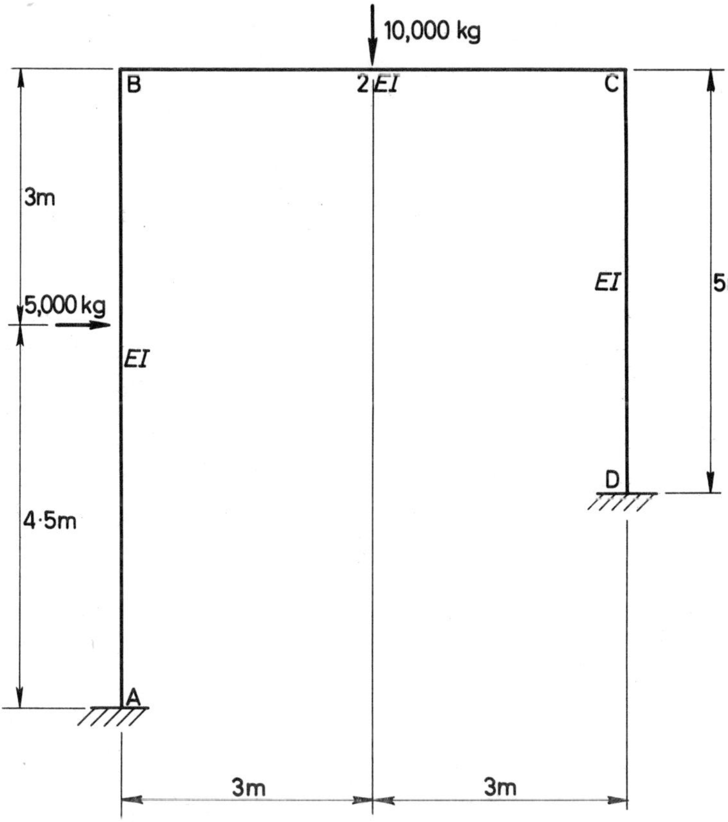

Figure 2.9. Problem 2.1

2.2. The bridge structure shown in Fig. 2.10 has hinged supports at C and D, a hinge at E and is supported with vertical restraint only at A and B. Joints F and G are rigid and all the members have the same uniform cross-section. Determine all the support reactions and the bending moments at F and G due to a vertical load of 40 kN at F.

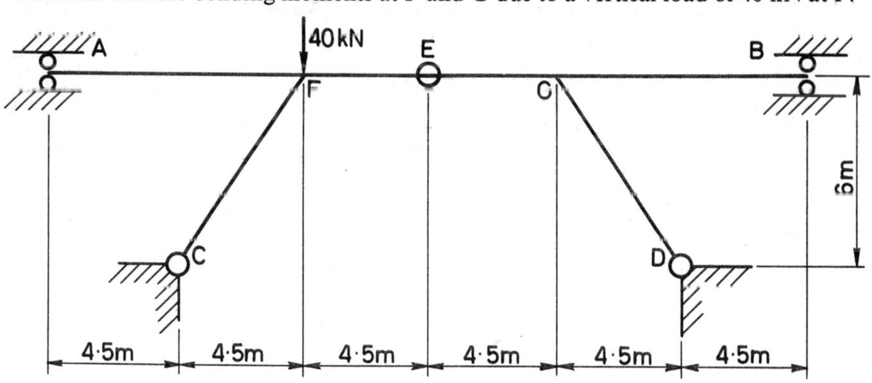

Figure 2.10. Problem 2.2

2.3. A Vierendeel girder, Fig. 2.11, is formed from two panels with rigid joints. Using the flexibility method, calculate the bending moments at D and E due to a central vertical load of W kg.

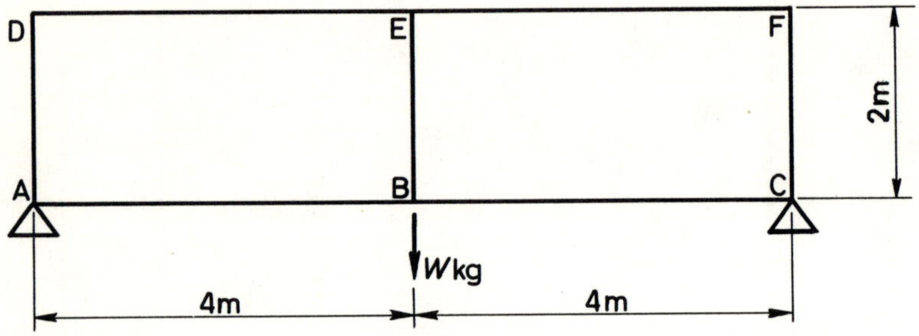

Figure 2.11. Problem 2.3

2.4. The frame shown in Fig. 2.12 is composed of members having the same flexural rigidity and carries a uniformly distributed load of 30 kN/m on the beam. The columns are fixed at support A and hinged at B. Calculate the bending moments at the joints and the deflections in the directions $R_{1,2,3,4}$. Use the complete matrix formulation of the flexibility method.

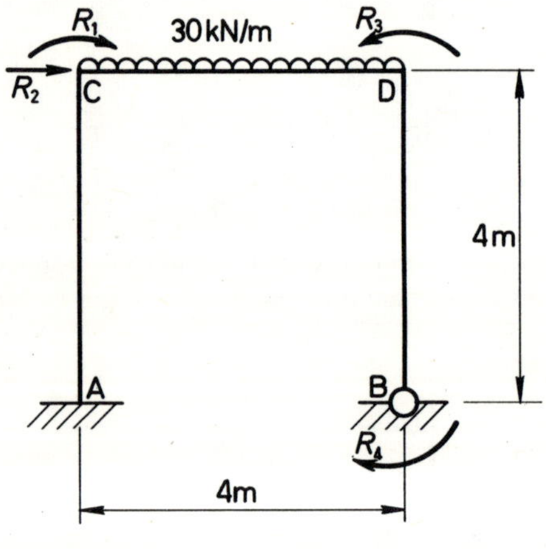

Figure 2.12. Problem 2.4

2.5. Show that the degree of statical indeterminacy of the frame shown in Fig. 2.13a is 9, and assemble a flexibility matrix for a primary structure as shown in Fig. 2.13b. The flexural rigidity of the beams is twice that of the columns.

2.6. The pin-jointed truss shown in Fig. 2.14 is hinged to a rigid support at A and stayed by cables BC, BD, BE. All three cables have L/AE equal to four times that of the members of the truss. All the truss members are of length l. Calculate the tensions in the cables.

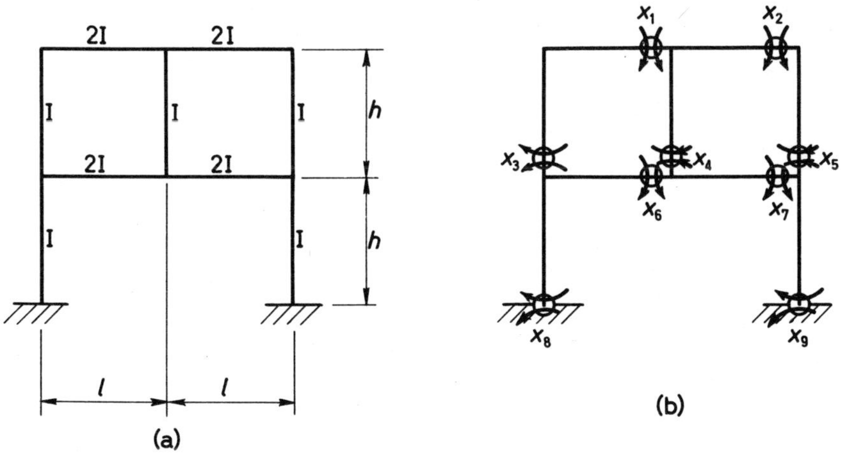

Figure 2.13. Problem 2.5

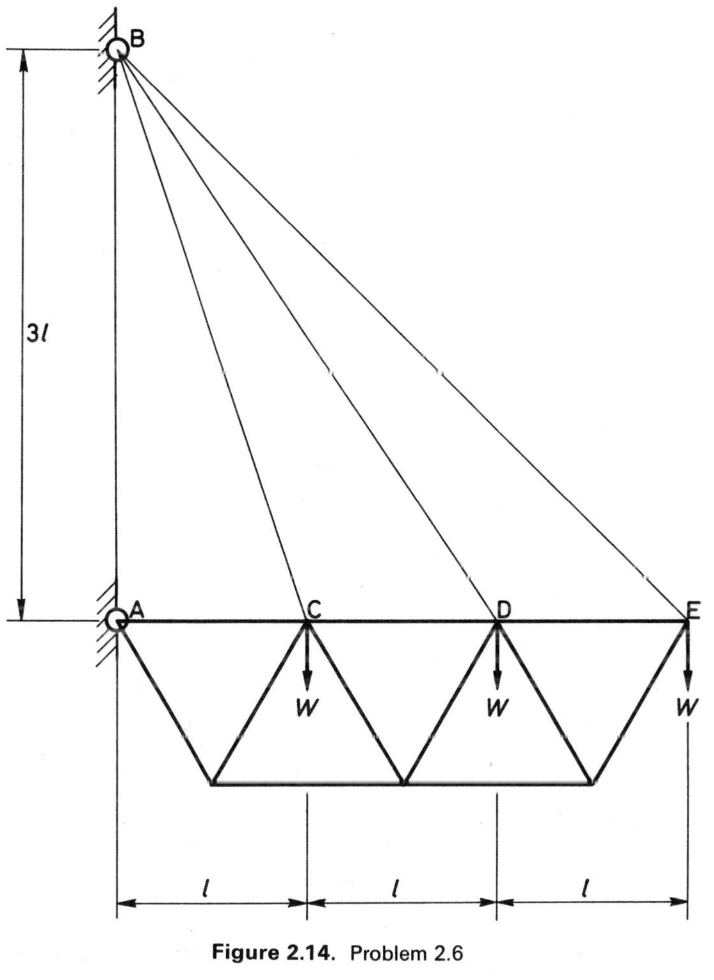

Figure 2.14. Problem 2.6

3. Additional Topics in the Flexibility Method

Following on the general outline of the flexibility method given in chapter 2, we now consider some further applications of the method to different types of structure and introduce certain refinements.

3.1. End Rotations of Non-prismatic Members

For the non-prismatic beam shown in Fig. 3.1, the end flexibilities are, from eq. (1.20),

$$\left.\begin{aligned} f_{11} &= \int_0^l \frac{m_1{}^2\,\mathrm{d}x}{EI} \\ f_{22} &= \int_0^l \frac{m_2{}^2\,\mathrm{d}x}{EI} \\ f_{12} = f_{21} &= \int_0^l \frac{m_1 m_2\,\mathrm{d}x}{EI} \end{aligned}\right\} \qquad (3.1)$$

If the EI variation along the beam, as a function of x, is such that the integrations in eq. (3.1) can be performed by direct methods then the calculation of the influence coefficients is straightforward. If however, as is often the case, the EI variation is such that direct methods are not possible then a numerical process is required. The most suitable numerical method, whether for hand or machine computation, is one employing Simpson's rule. The member is divided into n segments of equal length h, where n is even, and values of $m_r m_s/EI = \phi$ are calculated at the $(n + 1)$ stations, then

$$f_{rs} = h/3[\phi_1 + 4\phi_2 + 2\phi_3 + 4\phi_4 + \cdots + 2\phi_{n-1} + 4\phi_n + \phi_{n+1}]$$

$$= h/3\left[\begin{aligned} &\text{sum of end ordinates} + 4\,(\text{even ordinates}) \\ &\qquad\qquad + 2\,(\text{other odd ordinates}) \end{aligned}\right] \qquad (3.2)$$

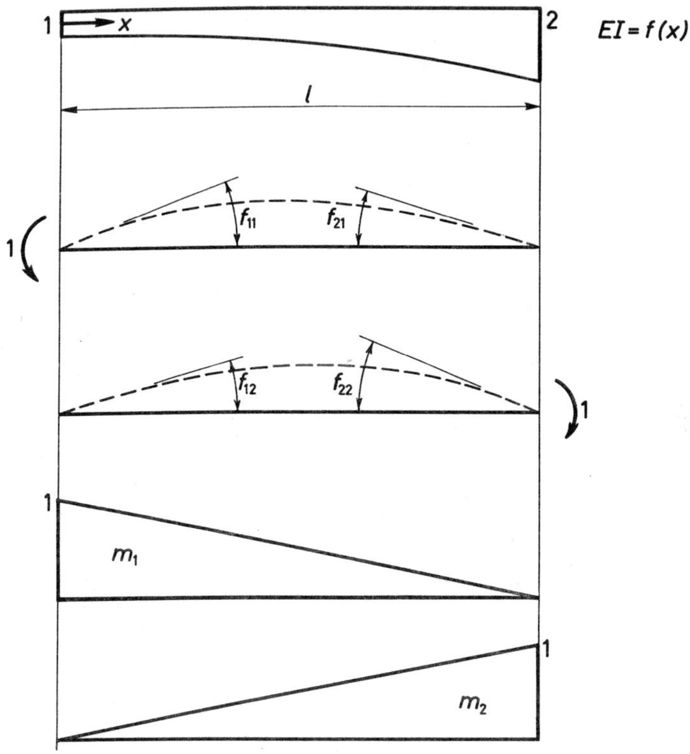

Figure 3.1. End flexibilities for non-prismatic member

Example 3.1. The beam shown in Fig. 3.2 has a stepped variation in *EI*. It is required to determine expressions for the end flexibilities. It will be sufficient in this case to divide each of the three parts of the beam into two segments and apply Simpson's rule separately to the parts.

$$f_{AA} = \frac{al}{6EI}\left\{\alpha + 4\alpha\frac{[l - (al/2)]^2}{l^2} + \alpha\frac{(l - al)^2}{l^2}\right\}$$

$$+ \frac{cl}{6EI}\left\{\frac{(l - al)^2}{l^2} + 4\frac{[l - al - (cl/2)]^2}{l^2} + b^2\right\}$$

$$+ \frac{bl}{6EI}\left\{b^2\beta + 4\frac{b^2}{4}\beta\right\}$$

$$= \frac{l}{3EI}\left\{\alpha a\left[a^2 + 3(1 - a)\right] + c\left[c^2 + 3b(1 - a)\right] + \beta b^3\right\} \qquad (3.3)$$

Similarly,

$$f_{BB} = \frac{l}{3EI}\{\beta b[b^2 + 3(1 - b)] + c[c^2 + 3a(1 - b)] + \alpha a^3\} \tag{3.4}$$

and

$$f_{AB} = f_{BA} = \frac{l}{6EI}\{\alpha a^2(3 - 2a) + c[1 - (a - b)^2 + a(1 - a) + b(1 - b)] + \beta b^2(3 - 2b)\} \tag{3.5}$$

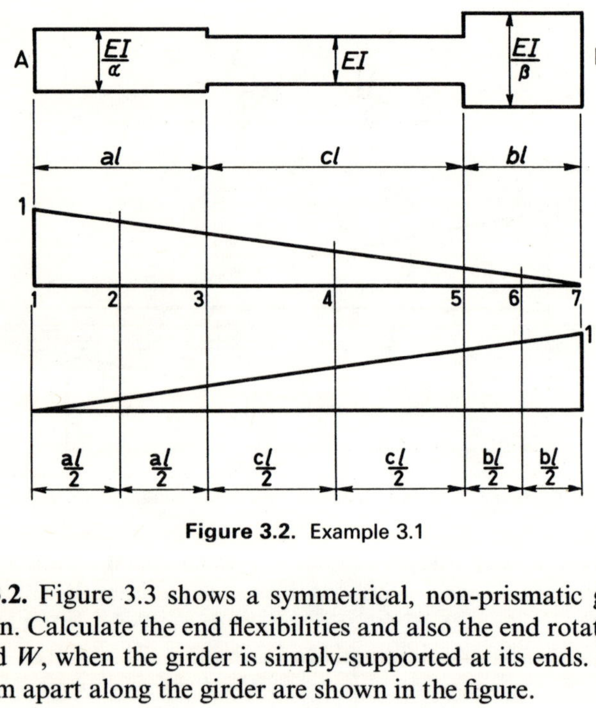

Figure 3.2. Example 3.1

Example 3.2. Figure 3.3 shows a symmetrical, non-prismatic girder of 30 metres span. Calculate the end flexibilities and also the end rotations caused by the load W, when the girder is simply-supported at its ends. EI values at stations 3 m apart along the girder are shown in the figure.

The data required for the numerical integration is set out in the following table:

x	0 (A)	3	6	9	12	15	18	21	24	27	30
$10^{-9}EI$ (kN mm² units)	1600	882	503	312	225	200	225	312	503	882	1600
m_1	1	0·9	0·8	0·7	0·6	0·5	0·4	0·3	0·2	0·1	0
m_2	0	0·1	0·2	0·3	0·4	0·5	0·6	0·7	0·8	0·9	1
m_0/W	0	2·4	4·8	4·2	3·6	3·0	2·4	1·8	1·2	0·6	0

Using Simpson's rule, eq. (3.2), to perform the integrations,

$$f_{AA} = f_{BB} = \int_0^{30} \frac{m_1{}^2\,dx}{EI} = \int_0^{30} \frac{m_2{}^2\,dx}{EI} = 24{\cdot}11 \times 10^{-6}$$

$$f_{AB} = f_{BA} = \int_0^{30} \frac{m_1 m_2\,dx}{EI} = 16{\cdot}74 \times 10^{-6}$$

$$\left.\begin{aligned} u_A &= \int_0^{30} \frac{m_1 m_0\,dx}{EI} = 128{\cdot}6 \times 10^{-6} W \\ u_B &= \int_0^{30} \frac{m_2 m_0\,dx}{EI} = 99{\cdot}08 \times 10^{-6} W \end{aligned}\right\} \text{(WkN)}$$

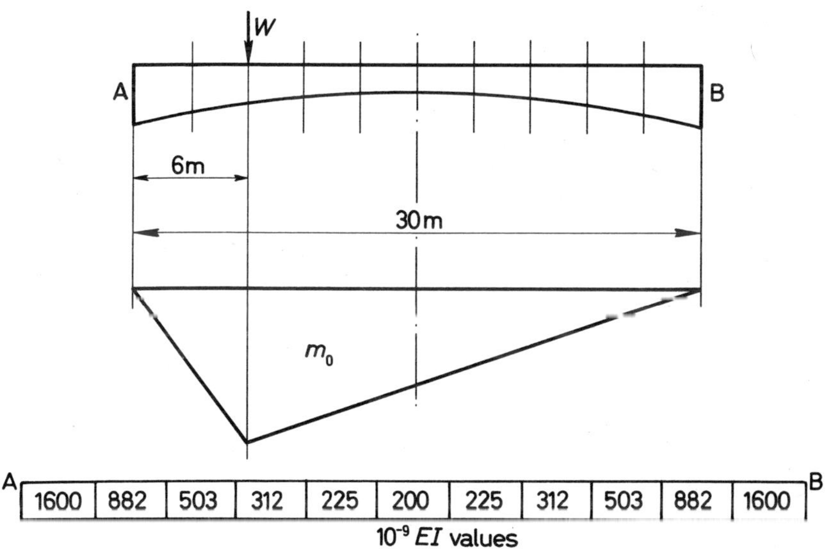

Figure 3.3. Example 3.2

Accuracy using Simpson's rule. For polynomials up to the third degree, Simpson's rule gives an exact evaluation of the integrals. In other cases, the accuracy of the result will depend on the number of ordinates taken. If a discontinuity occurs in the *EI* variation, or in one of the bending moment diagrams, at an even-numbered station, this will reduce the accuracy of the result. This effect can be overcome by halving the intervals, thus causing the discontinuity to fall at an odd-numbered station.

3.2. Multi-action Strain Energies

In the problems examined so far, flexural deformations have predominated and flexibilities have been calculated on the basis of flexural strain energy according to equations of type (1.20). In certain circumstances it may be necessary to include deformations resulting from other actions, and a more general form of energy expression may be required, for example,

$$U = \int \frac{M^2\,ds}{2EI} + \int \frac{N^2\,ds}{2EA} + f \int \frac{V^2\,ds}{2AG} \tag{3.6}$$

In eq. (3.6), M, N, and V represent distributions of bending moment, thrust, and shear respectively, and f is a shape constant for the cross-section. An expression for f is given in eq. (3.16), and values of f for some common cross-sections are included in Table 3.1. In many cases, some terms are zero or negligible. Taking the first two terms in eq. (3.6), and using Castigliano's theorem, eq. (1.19),

$$f_{rs} = \int \frac{m_r m_s\,ds}{EI} + \int \frac{n_r n_s\,ds}{EA} \tag{3.7}$$

$$u_r = \int \frac{m_0 m_r\,ds}{EI} + \int \frac{n_0 n_r\,ds}{EA} \tag{3.8}$$

Thus the influence coefficient expressions now contain two terms and two sets of diagrams are required. Otherwise, the procedure is the same as for the single energy case.

Effects of axial forces in the flexibility method

Example 3.3. The portal frame shown in Fig. 3.4 is to be analysed for the three alternative positions of a single 100 kN load. Axial strain energy is to be included and all the members are taken to have the same uniform cross-section, with $A = 3I_0$ (metre units). The influence coefficients, calculated

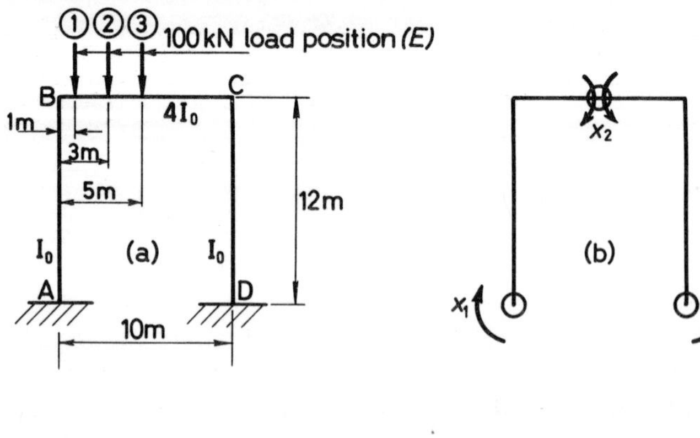

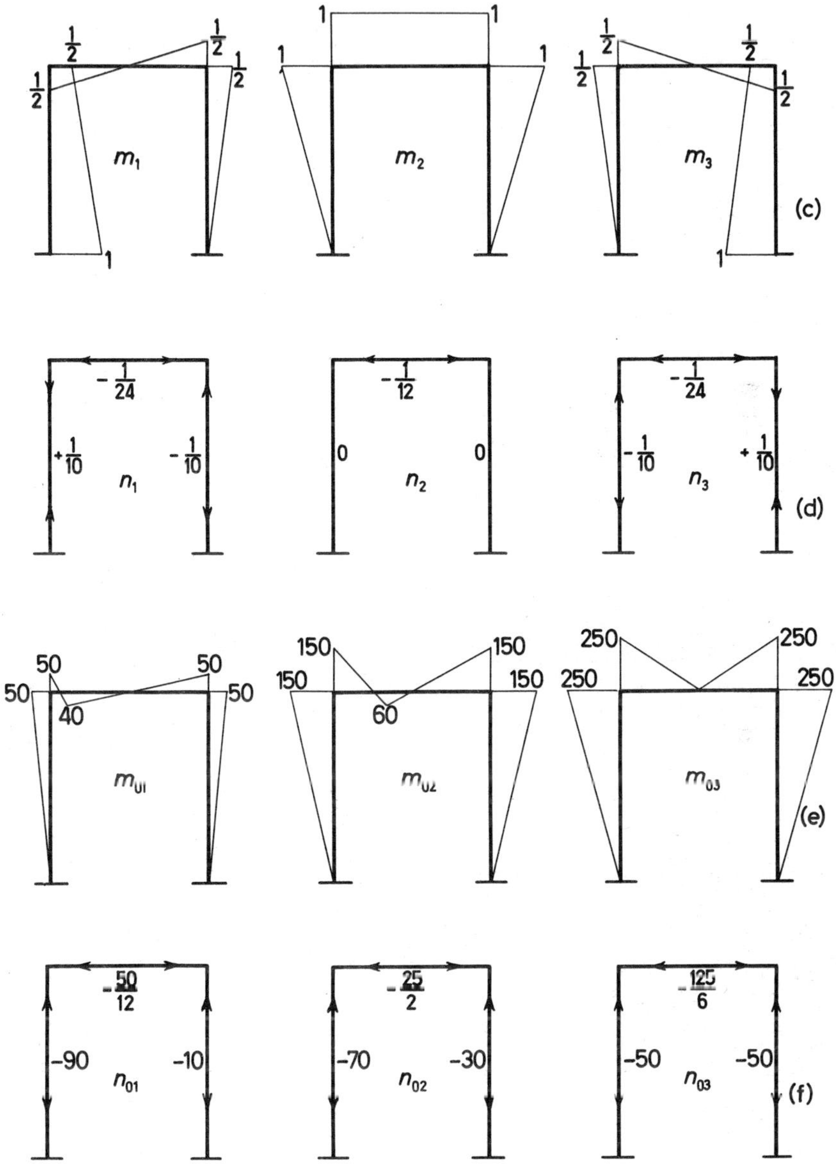

Figure 3.4. Example 3.3

using the table of product integrals, Appendix A, and the unit diagrams in Fig. 3.4, are set out in the table overleaf. (EI_0 may be omitted from the calculation since constant for the frame members.)

								Load position 1			2			3		
	Member	f_{11}	f_{22}	f_{33}	f_{12}	f_{23}	f_{13}	u_1	u_2	u_3	u_1	u_2	u_3	u_1	u_2	u_3
Bending	AB	7·0	4·0	1·0	−4·0	2·0	−2·0	−200	200	100	−600	600	300	−1000	1000	500
	BE	0·208	2·5	0·208	0	0	−0·104	15	−18·75	−15	17·5	112·5	−17·5	0	312·5	0
	EC						−0·104		31·25							
	CD	1·0	4·0	7·0	2·0	−4·0	−2·0	100	200	−200	300	600	−600	500	1000	−1000
	Totals	8·21	10·5	8·21	−2·0	−2·0	−4·21	−85	412·5	−115	−282·5	1312·5	−317·5	−500	2312·5	−500
Thrust	AB	0·04	0	0·04	0	0	−0·04	−36	0	36	−28	0	28	−20	0	20
	BC	0·006	0·024	0·006	0·012	0·012	0·006	0·58	1·16	0·58	1·73	3·46	1·73	2·9	5·8	2·9
	CD	0·04	0	0·04	0	0	−0·04	4	0	−4	12·0	0	−12·0	20	0	−20
	Totals	0·086	0·024	0·086	0·012	0·012	−0·074	−31·42	1·16	32·58	−14·3	3·46	17·7	2·9	5·8	2·9

Considering bending energy alone,

$$\mathbf{F}_b = \begin{bmatrix} 8{\cdot}21 & -2{\cdot}00 & -4{\cdot}21 \\ -2{\cdot}00 & 10{\cdot}50 & -2{\cdot}00 \\ -4{\cdot}21 & -2{\cdot}00 & 8{\cdot}21 \end{bmatrix} \tag{3.9}$$

and

$$\mathbf{F}_b^{-1} = \begin{bmatrix} 0{\cdot}196 & 0{\cdot}0595 & 0{\cdot}115 \\ 0{\cdot}0595 & 0{\cdot}118 & 0{\cdot}0595 \\ 0{\cdot}115 & 0{\cdot}0595 & 0{\cdot}196 \end{bmatrix} \tag{3.10}$$

For bending and axial energies combined,

$$\mathbf{F}_{bd} = \begin{bmatrix} 8{\cdot}30 & -1{\cdot}99 & -4{\cdot}28 \\ -1{\cdot}99 & 10{\cdot}52 & -1{\cdot}99 \\ -4{\cdot}28 & -1{\cdot}99 & 8{\cdot}30 \end{bmatrix} \tag{3.11}$$

$$\mathbf{F}_{bd}^{-1} = \begin{bmatrix} 0{\cdot}194 & 0{\cdot}0581 & 0{\cdot}115 \\ 0{\cdot}0581 & 0{\cdot}117 & 0{\cdot}0581 \\ 0{\cdot}115 & 0{\cdot}0581 & 0{\cdot}194 \end{bmatrix} \tag{3.12}$$

The values of the redundants (kNm units) corresponding to the three alternative load positions are, for bending only,

$$\mathbf{X}_b = -\mathbf{F}_b^{-1} \begin{bmatrix} -85{\cdot}0 & -282{\cdot}5 & -500 \\ 412{\cdot}5 & 1312{\cdot}5 & 2312{\cdot}5 \\ -115{\cdot}0 & -317{\cdot}5 & -500 \end{bmatrix}$$

$$= \begin{bmatrix} 5{\cdot}3 & 13{\cdot}9 & 18{\cdot}0 \\ -36{\cdot}7 & -119{\cdot}3 & -213{\cdot}4 \\ 7{\cdot}7 & 16{\cdot}7 & 18{\cdot}0 \end{bmatrix} \tag{3.13}$$

and for bending and thrust combined

$$\mathbf{X}_{bd} = -\mathbf{F}_{bd}^{-1} \begin{bmatrix} -116{\cdot}4 & -296{\cdot}8 & -497{\cdot}1 \\ 413{\cdot}7 & 1316 & 2318 \\ -82{\cdot}4 & -300 & -497{\cdot}1 \end{bmatrix}$$

$$= \begin{bmatrix} 7{\cdot}3 & 15{\cdot}6 & 18{\cdot}9 \\ -37{\cdot}1 & -119{\cdot}3 & -213{\cdot}5 \\ 4{\cdot}4 & 15{\cdot}9 & 18{\cdot}9 \end{bmatrix} \tag{3.14}$$

The effect of inclusion of the axial deformations of the members can be assessed by comparing the matrices (3.13) and (3.14).

Effects of shear forces in the flexibility method. The addition of a term to account for energy due to shearing deformations, in the strain energy expression (3.6), has the effect of increasing the calculated deflections. As is well known in the theory of beams, the effect of shear on the beam deflections is negligible unless the beam has a small span/depth ratio. In general, shear does not greatly affect the member flexibilities, but cases do arise where it is necessary to take shear into account and we shall now consider a suitable method.

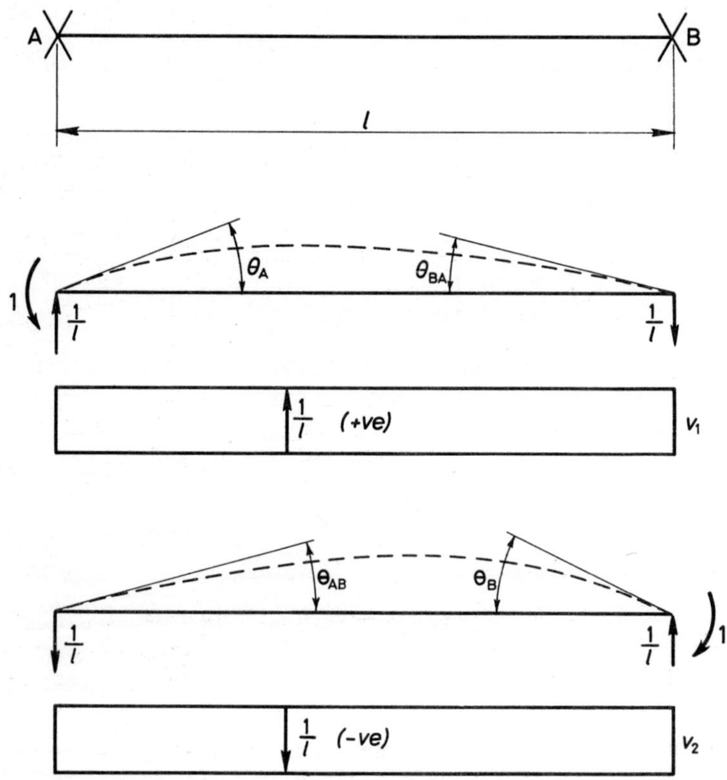

Figure 3.5

In the calculation of the end flexibilities of the beam AB, Fig. 3.5, the influence of shearing deformations can be included as follows:

$$f_{11} = \theta_A = \int_0^l \frac{m_1{}^2\,dx}{EI} + f\int_0^l \frac{v_1{}^2\,dx}{GA} \tag{3.15}$$

in which v_1 is the shear force distribution due to unit moment at A, and f is a dimensionless form factor, dependent only on the shape of the cross-section.

$$f = \frac{A}{I^2}\int \frac{Q^2\,\mathrm{d}A}{b^2} \tag{3.16}$$

in which, A = cross-sectional area
b = width of section
Q = first moment of area of the part of the cross-section above the section considered, about the neutral axis.

Values of the form factor f, for some common cross-sections, are given in Table 3.1.

Table 3.1. Form factors for calculating deflections due to shear

Rectangle	1·2
Circle	10/9
Hollow circle	2
I-section or hollow rectangle	A/A_{web} approximately

If the beam AB in Fig. 3.5 has a uniform cross-section, then eq. (3.15) becomes

$$f_{11} = l/3EI + f/lGA = f_{22} \tag{3.17}$$

and

$$f_{12} = f_{21} = l/6EI - f/lGA \tag{3.18}$$

3.3. Member Curved in the Plane of the Applied Loading

The calculation of flexibilities for curved elements, such as arches and rings, may be carried out by the foregoing methods, keeping in mind that the integrations must be carried out along the length of the element. Consider a circular arc segment AB, Fig. 3.6, and neglect axial deformations. The bending moment distributions due to unit moment at A, unit moment at B, and unit horizontal thrust are, respectively,

$$\left.\begin{aligned} m_1 &= 1 - \frac{R(\sin\alpha + \sin\theta)}{2R\sin\alpha} = \frac{1}{2}\left(1 - \frac{\sin\theta}{\sin\alpha}\right) \\ m_2 &= \frac{R(\sin\alpha + \sin\theta)}{2R\sin\alpha} = \frac{1}{2}\left(1 + \frac{\sin\theta}{\sin\alpha}\right) \\ m_3 &= -R(\cos\theta - \cos\alpha) = R(\cos\alpha - \cos\theta) \end{aligned}\right\} \tag{3.19}$$

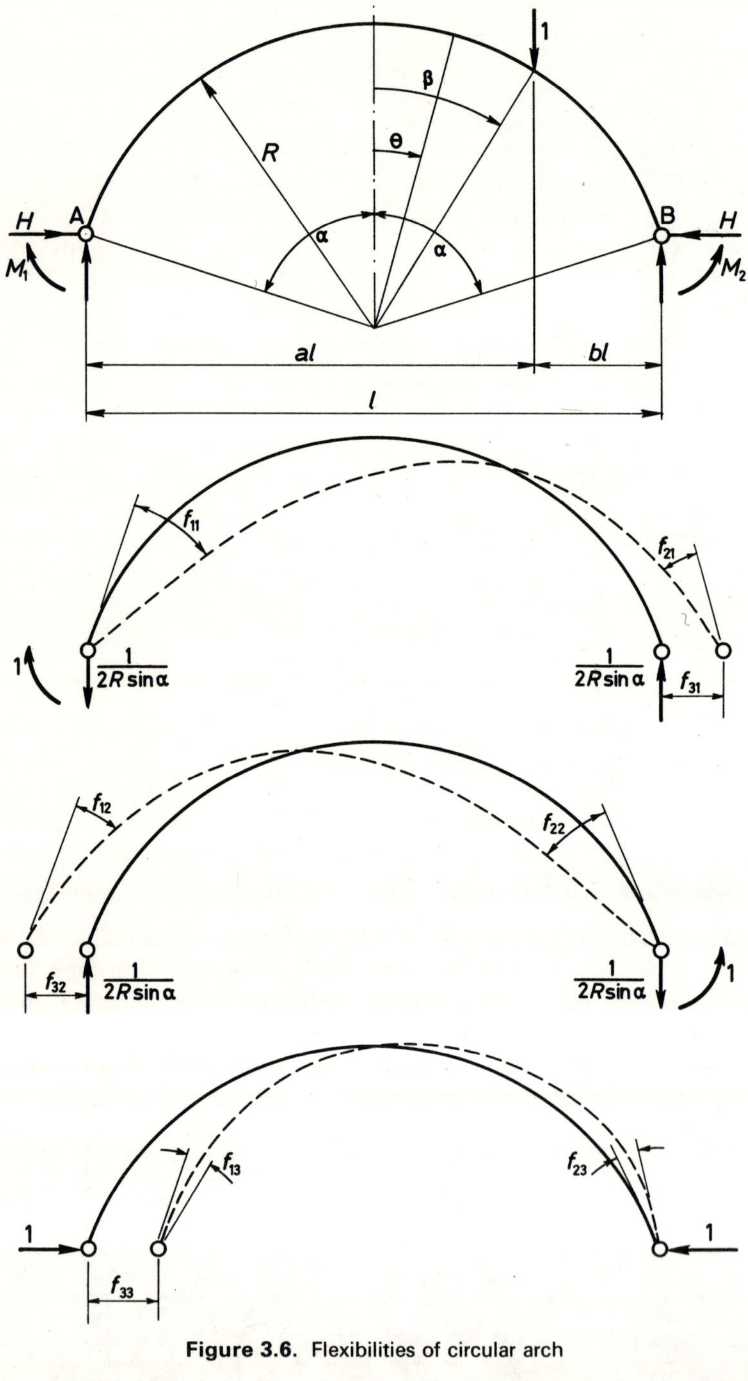

Figure 3.6. Flexibilities of circular arch

Thus

$$f_{11} = \int_{-\alpha}^{\alpha} \frac{m_1{}^2\,ds}{EI} = \int_{\alpha}^{-\alpha} \frac{m_1{}^2}{EI} R\,d\theta = f_{22}$$

$$= \frac{R}{8EI\sin^2\alpha}(2\alpha + 4\alpha\sin^2\alpha - \sin 2\alpha)$$

$$f_{33} = \int_{-\alpha}^{\alpha} \frac{m_3{}^2\,ds}{EI} = \frac{R^3}{2EI}(2\alpha + 4\alpha\cos^2\alpha - 3\sin 2\alpha)$$

$$f_{12} = f_{21} = \int_{\alpha}^{-\alpha} \frac{m_1 m_2\,ds}{EI} = \frac{R}{8EI\sin^2\alpha}(-2\alpha + \sin 2\alpha + 4\alpha\sin^2\alpha)$$

$$f_{13} = f_{31} = \int_{\alpha}^{-\alpha} \frac{m_1 m_3\,ds}{EI} = \frac{R^2}{EI}(\alpha\cos\alpha - \sin\alpha)$$

$$f_{23} = f_{32} = \int_{\alpha}^{-\alpha} \frac{m_2 m_3\,ds}{EI} = \frac{R^2}{EI}(\alpha\cos\alpha - \sin\alpha)$$

Thus, the flexibility matrix for the arch rib is

$$\mathbf{F} = \frac{R}{8EI\sin^2\alpha}\begin{bmatrix} 2\alpha + 4\alpha\sin^2\alpha - \sin 2\alpha & & \text{Symmetric} \\ 4\alpha\sin^2\alpha - 2\alpha + \sin 2\alpha & 2\alpha + 4\alpha\sin^2\alpha - \sin 2\alpha & \\ 8R\sin^2\alpha(\alpha\cos\alpha - \sin\alpha) & 8R\sin^2\alpha(\alpha\cos\alpha - \sin\alpha) & 4R^2\sin^2\alpha(2\alpha + 4\alpha\cos^2\alpha - 3\sin 2\alpha) \end{bmatrix} \quad (3.20)$$

As an illustration of the calculation of displacements at A and B in the directions of the redundants, consider a unit point load applied to the arch at a position defined by β, Fig. 3.6.

The vertical reactions at A and B are b and a respectively. The statically determinate bending moment distribution due to the applied load is given by

$$\left.\begin{aligned} m_0 &= bR(\sin\alpha + \sin\theta) \quad \text{for } -\alpha \leqslant \theta \leqslant \beta \\ m_0 &= aR(\sin\alpha - \sin\theta) \quad \text{for } \beta \leqslant \theta \leqslant \alpha \end{aligned}\right\} \quad (3.21)$$

Thus

$$u_1 = \int_{-\alpha}^{\alpha} \frac{m_0 m_1}{EI} R\,d\theta$$

$$= \frac{R^3}{4EIl}\{4[\alpha - \beta(2a - 1)]\sin^2\alpha + (1 + 2a)\sin 2\alpha - 2[\beta - \alpha(2a - 1)] + 2\cos\beta(\sin\beta - 4a\sin\alpha)\} \quad (3.22)$$

Table 3.2. Circular Arch Flexibilities and End Displacements due to Unit Load

			α°							
			20	30	40	50	60	70	80	90
$\frac{EI}{R} f_{11} = \frac{EI}{R} f_{22}$			0·2337	0·3524	0·4735	0·5983	0·7283	0·8658	1·0140	1·1781
$\frac{EI}{R} f_{12}$			0·1154	0·1712	0·2246	0·2743	0·3189	0·3560	0·3823	0·3927
$\frac{EI}{R^2} f_{13} = \frac{EI}{R^2} f_{23}$			−0·0140	−0·0466	−0·1080	−0·2051	−0·3424	−0·5218	−0·7423	−1·0000
$\frac{EI}{R^3} f_{33}$			0·00135	0·00996	0·04028	0·1166	0·2718	0·5434	0·9674	1·5708
$\frac{EI}{R^2} u_1$	a	0·1	0·0136	0·02977	0·05087	0·07534	0·1011	0·1253	0·1442	0·1521
		0·2	0·0228	0·04969	0·08429	0·1236	0·1638	0·2001	0·2270	0·2376
		0·3	0·02820	0·06121	0·1033	0·1506	0·1981	0·2402	0·2706	0·2823
		0·4	0·03028	0·06558	0·1104	0·1603	0·2100	0·2535	0·2844	0·2962
		0·5	0·02954	0·06391	0·1074	0·1556	0·2035	0·2450	0·2743	0·2854

$\frac{EI}{R^2}u_2$	a	0·1	0·00781	0·01693	0·02853	0·04149	0·0544	0·06576	0·0738	0·0768
		0·2	0·01514	0·03279	0·05516	0·08006	0·1041	0·1263	0·1414	0·1471
		0·3	0·02151	0·04654	0·07822	0·1134	0·1481	0·1783	0·1995	0·2074
		0·4	0·02646	0·05724	0·09614	0·1393	0·1819	0·2189	0·2448	0·2546
		0·5	0·02954	0·06391	0·1074	0·1556	0·2035	0·2450	0·2743	0·2854
$\frac{EI}{R^3}u_3$	a	0·1	−0·00094	−0·00455	−0·01355	−0·03054	−0·05725	−0·09358	−0·1367	−0·1800
		0·2	−0·00177	−0·00857	−0·02539	−0·05687	−0·1057	−0·1709	−0·2464	−0·3200
		0·3	−0·00242	−0·01168	−0·03445	−0·07679	−0·1419	−0·2277	−0·3257	−0·4200
		0·4	−0·00283	−0·01363	−0·04012	−0·08914	−0·1641	−0·2623	−0·3736	−0·4800
		0·5	−0·00297	−0·01430	−0·04205	−0·09332	−0·1716	−0·2739	−0·3896	−0·5000

and $$u_2 = \int_{-\alpha}^{\alpha} \frac{m_0 m_2}{EI} R\, d\theta$$

$$= \frac{R^3}{4EIl} \{4[\alpha - \beta(2a - 1)] \sin^2 \alpha + (1 + 2b) \sin 2\alpha$$

$$+ 2[\beta + \alpha(2b - 1)] - 2 \cos \beta(\sin \beta + 4b \sin \alpha)\} \quad (3.23)$$

and $$u_3 = \int_{-\alpha}^{\alpha} \frac{m_0 m_3}{EI} R\, d\theta$$

$$= \frac{R^3}{4EI} \{2[\alpha - \beta(2a - 1)] \sin 2\alpha + 4(2a - 1) \sin \alpha \sin \beta$$

$$- 4 \cos \alpha \cos \beta + \cos 2\beta + 3 \cos 2\alpha\} \quad (3.24)$$

The circular arch flexibilities given by eq. (3.20) and the end displacements given by eqns. (3.22), (3.23), and (3.24) are tabulated in Table 3.2 over a practical range of values of the parameters.

Example 3.4. A circular segmental arch of 100 metres span is shown in Fig. 3.7. It is required to determine values of the redundant reactions due to the three concentrated loads W shown.

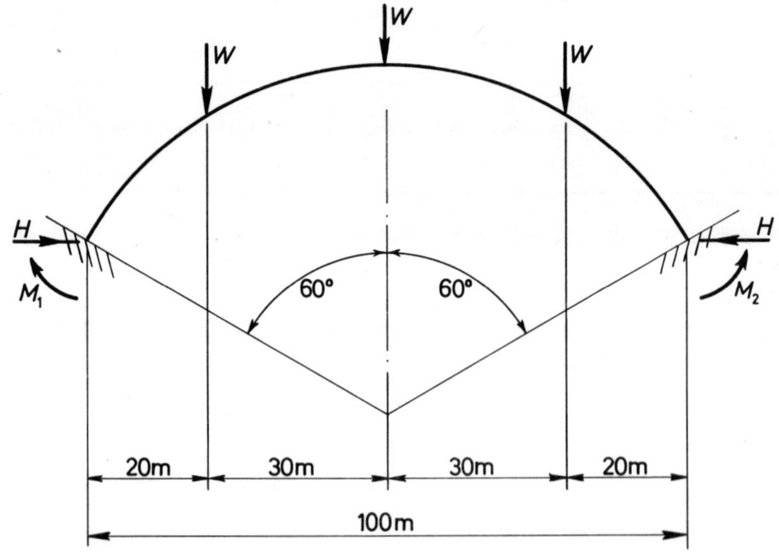

Figure 3.7. Example 3.4

The flexibilities are obtained from Table 3.2, using $\alpha = 60°$ and $R = 100/\sqrt{3}$.

$\mathbf{F} = 1/EI$

$72{\cdot}83/\sqrt{3}$		Symmetric
$31{\cdot}89/\sqrt{3}$	$72{\cdot}83/\sqrt{3}$	
$-3424/3$	$-3424/3$	$271\,800/3\sqrt{3}$

$= 1/EI$

42·1		Symmetric
18·4	42·1	
−1140	−1140	52 300

$$u_1 = \frac{100^2 W}{3EI}\{0{\cdot}1638 + 0{\cdot}2035 + 0{\cdot}1041\} = 1571\frac{W}{EI} = u_2$$

$$u_3 = \frac{100^3 W}{3\sqrt{3}EI}\{-0{\cdot}1057 \times 2 - 0{\cdot}1716\} = -73\,700\frac{W}{EI}$$

$$\mathbf{X} = -\mathbf{F}^{-1}\mathbf{U} = \begin{bmatrix} M_1 \\ M_2 \\ H \end{bmatrix} = W\begin{bmatrix} 3{\cdot}62 \\ 3{\cdot}62 \\ 1{\cdot}57 \end{bmatrix}$$

3.4. Combined Bending and Torsion

Examples of structures in which torsional effects are of importance, are the beam curved in a plane not coincident with the plane of the applied loading, and the rigid frame loaded perpendicular to its plane. The effect of torsion can be included in the flexibility method by observing the relationship between twisting moment M_x and rotation ϕ of a uniform member, of length l,

$$M_x = GJ\phi/l \tag{3.25}$$

where GJ = torsional rigidity of the member.

Thus, the end flexibility for torsion is

$$f_T = l/GJ \tag{3.26}$$

Some values of J for structural sections will be found in Appendix C. Thus a flexibility matrix, suitable for a straight, uniform member, including the effect of twisting is, eq. (3.27),

$$\mathbf{F} = l/6EI \begin{bmatrix} 2 & -1 & 0 \\ -1 & 2 & 0 \\ 0 & 0 & 6\alpha \end{bmatrix} \tag{3.27}$$

where $$\alpha = EI/GJ \tag{3.28}$$

Example 3.5. A rigidly-jointed frame ABCB′A′ lies in a horizontal plane and is encastré at A and A′, Fig. 3.8. A vertical load P is carried at C, the mid-point of BB′. All members have the same uniform cross-section, with a

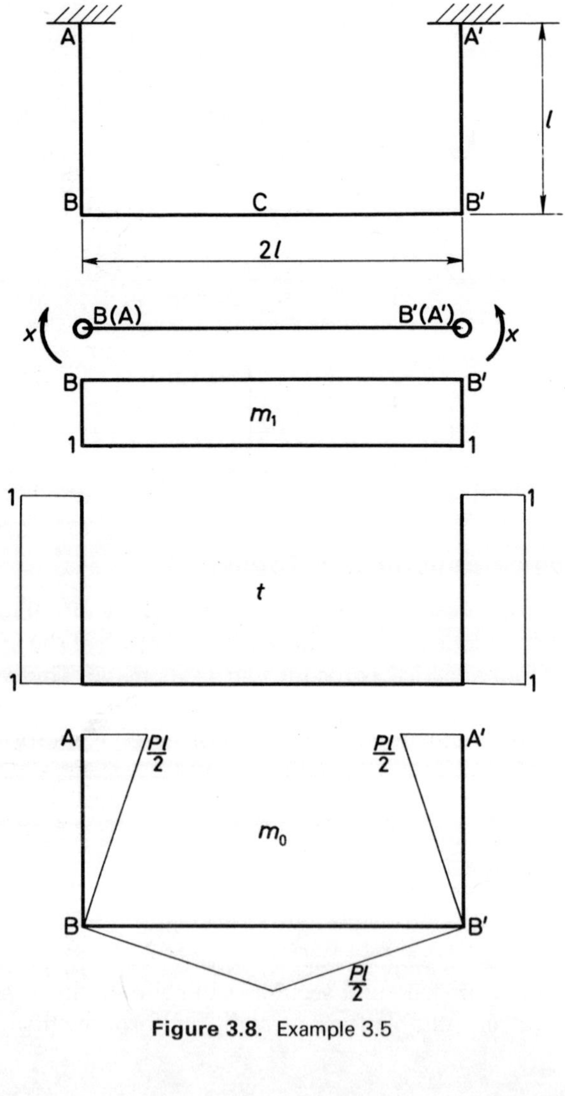

Figure 3.8. Example 3.5

flexural rigidity EI and a torsional rigidity GJ. Due to the symmetry of the structure and the loading there is only one redundancy, which may be taken to be the end bending moments on BB′, which are transmitted to BA and B′A′ as twisting moments. The bending and twisting moment diagrams in the statically determinate condition are as shown in Fig. 3.8.

$$f_{11} = \int \frac{m_1^2 \, ds}{EI} + \int \frac{t^2 \, ds}{GJ} = \frac{2l}{EI} + \frac{2l}{GJ} = 2l\left(\frac{1}{EI} + \frac{1}{GJ}\right)$$

$$u = \int \frac{m_1 m_0 \, ds}{EI} + \int \frac{t \, t_0 \, ds}{GJ} = \frac{2l}{2} \times \frac{1}{EI} \times \frac{Pl}{2} + 0 = \frac{Pl^2}{2EI}$$

$$f_{11}x + u = 0$$

$$x = -\frac{Pl}{4}\left(\frac{GJ}{GJ + EI}\right)$$

Thus, the bending moment at C

$$= (m_0 + m_1 x)_C$$

$$= \frac{Pl}{2} - \frac{Pl}{4}\left(\frac{GJ}{GJ + EI}\right)$$

$$= \frac{Pl}{4}\left(\frac{GJ + 2EI}{GJ + EI}\right)$$

and the bending moment at A and A′

$$= (m_0 + m_1 x)_{A, A'}$$

$$= \frac{Pl}{2}$$

3.5. The Torsion-free Grillage

A grillage in which the member connections are such that only transverse forces can be transmitted, or in which the members have negligible torsional stiffness, is a 'torsion-free' grillage. The flexibility method is very suitable for the analysis of this type of grillage. The redundant forces are the forces transmitted at the nodes of the grillage and the compatibility conditions are obtained by equating the deflections of the members meeting at the nodes.

Example 3.6. A plan of a skew bridge is shown in Fig. 3.9. The bridge consists of three longitudinal girders AA, BB, AA and three cross-girders AFE, CDC, and EFA. The longitudinal girders are simply-supported at their ends AA, BB, AA and the connections between the longitudinal and cross girders are such that only vertical forces can be transmitted. All the girders have the same uniform cross-section. Determine the maximum bending moment in the cross-girder CC when the bridge is loaded with vertical loads W at F, D, F. A simply-supported beam of length l and flexural rigidity EI, subjected to a concentrated load W at a from one end, deflects $(Wac/6EI)(l^2 - a^2 - c^2)$ at c from the other end where $c \not> (l - a)$.

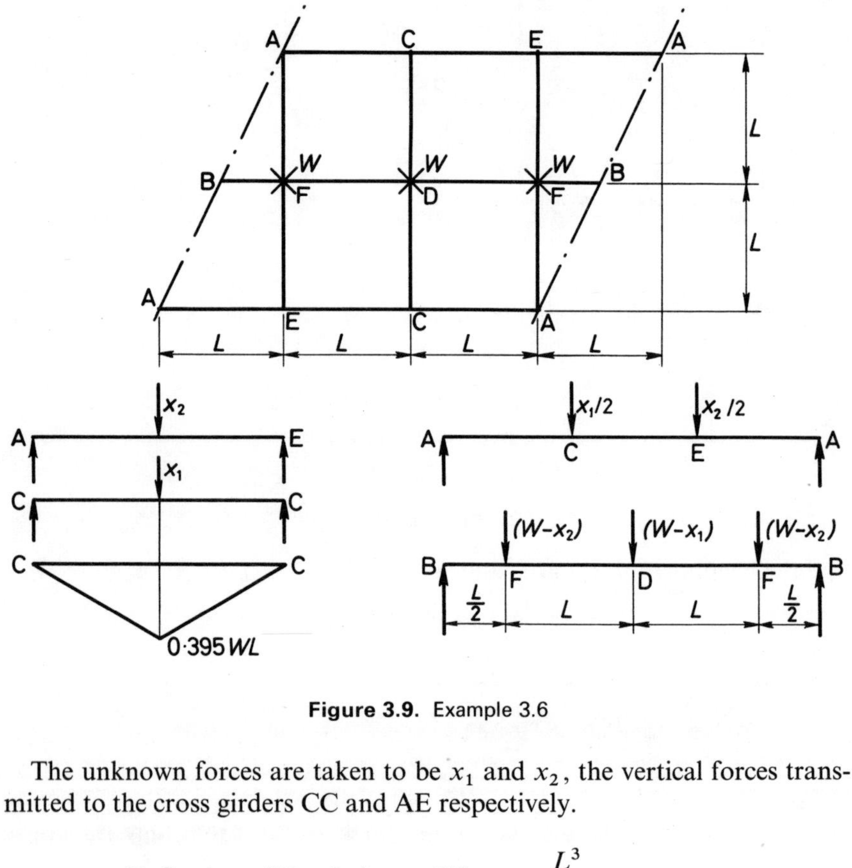

Figure 3.9. Example 3.6

The unknown forces are taken to be x_1 and x_2, the vertical forces transmitted to the cross girders CC and AE respectively.

$$\text{Deflection of D relative to CC} = x_1 \frac{L^3}{6EI}$$

$$\text{,, F ,, ,, AE} = x_2 \frac{L^3}{6EI}$$

$$\text{,, C ,, ,, AA} = \frac{2}{9} x_1 \frac{L^3}{EI} + \frac{7}{36} x_2 \frac{L^3}{EI}$$

$$\text{,, E ,, ,, AA} = \frac{2}{9} x_2 \frac{L^3}{EI} + \frac{7}{36} x_1 \frac{L^3}{EI}$$

Thus, the deflection of D $= \dfrac{x_1 L^3}{6EI} + \dfrac{2}{9} x_1 \dfrac{L^3}{EI} + \dfrac{7}{36} x_2 \dfrac{L^3}{EI}$

and the deflection of F $= \dfrac{x_2 L^3}{6EI} + \dfrac{1}{2}\left(\dfrac{2}{9} x_2 \dfrac{L^3}{EI} + \dfrac{7}{36} x_1 \dfrac{L^3}{EI}\right)$

Consider the longitudinal BB,

Deflection of D $= \dfrac{9}{16}\dfrac{L^3}{EI}(W - x_1) + \dfrac{13}{24}\dfrac{L^3}{EI}(W - x_2)$

Deflection of F $= \dfrac{42}{144}\dfrac{L^3}{EI}(W - x_2) + \dfrac{39}{144}\dfrac{L^3}{EI}(W - x_1)$

The compatibility conditions at D and F are obtained by equating the deflections of the longitudinal and cross girders, thus,

$$137x_1 + 106x_2 = 159W$$

$$53x_1 + 82x_2 = 81W$$

From which

$$x_1 = 0{\cdot}79W$$

$$x_2 = 0{\cdot}48W$$

The resulting bending moment diagram for the cross girder CC is shown in Fig. 3.9.

3.6. Space Truss Analysis

A method of analysis of plane trusses, using the flexibility method, was developed in section 2.7. We consider now the extension of this method to the analysis of trusses in space. The notation and basic equations developed in section 2.7 apply here and these are summarized in Table 3.3.

Example 3.7. In the pin-jointed space frame shown in Fig. 3.10, the members 3, 4, 5, 6, and 7 form a rectangular panel lying in a horizontal plane 6 metres above the pinned support E. All the members have the same extensional rigidity AE. The frame has one redundancy and will be analysed using the matrix scheme of Table 3.3. The force in member 7 is taken as the redundancy and the member forces obtained separately due to the applied loads and the redundant force. The member forces are determined from the conditions of statical equilibrium of the joints C and D in the directions X, Y, and Z. The

Table 3.3. Summary of flexibility method applied to truss analysis

	Statically determinate	Statically indeterminate
Element forces due to redundants		$\mathbf{Y} = \mathbf{TX}$
Element forces, in statically determinate condition, due to applied loads	$\mathbf{Y}_0 = \mathbf{T}_0\mathbf{R}$	$\mathbf{Y}_0 = \mathbf{T}_0\mathbf{R}$
Flexibility matrix (primary structure)		$\mathbf{F}_x = \mathbf{T}^T\mathbf{F}_y\mathbf{T}$
Primary structure displacements at positions of redundants		$\mathbf{U}_x = (\mathbf{T}^T\mathbf{F}_y\mathbf{T}_0)\mathbf{R}$
Flexibilities of unassembled members	$\mathbf{F}_y$	$\mathbf{F}_y$
Compatibility conditions		$\mathbf{F}_x\mathbf{X} + \mathbf{U} = \mathbf{0}$
Total element forces	$\mathbf{Y}_0 = \mathbf{T}_0\mathbf{R}$	$\mathbf{S} = \mathbf{T}_0\mathbf{R} + \mathbf{TX}$
Nodal displacements at applied load positions	$\mathbf{r} = (\mathbf{T}_0{}^T\mathbf{F}_y\mathbf{T}_0)\mathbf{R}$	$\mathbf{r} = \mathbf{T}_0{}^T\mathbf{F}_y(\mathbf{T}_0\mathbf{R} + \mathbf{TX})$

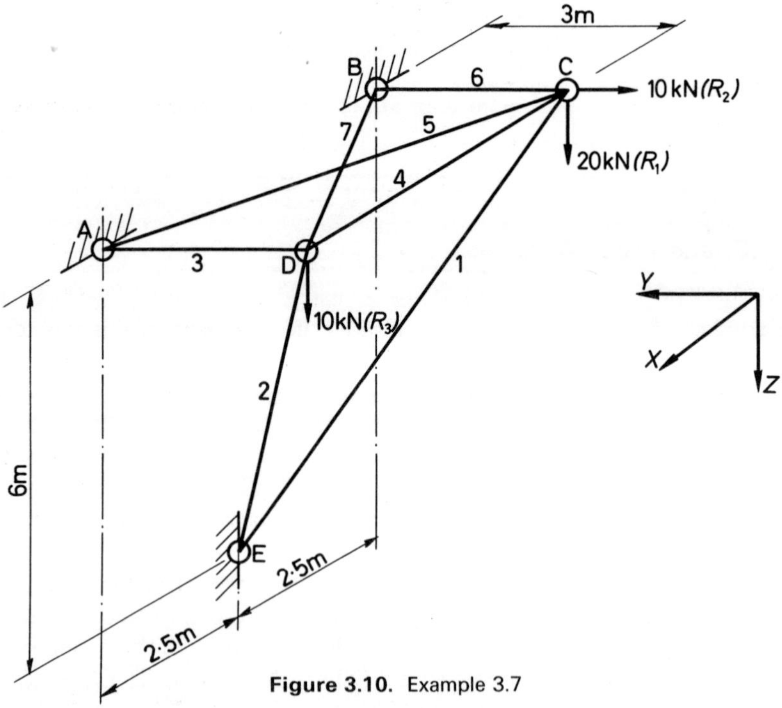

Figure 3.10. Example 3.7

method of tension coefficients is suitable for this purpose. Thus,

$$\begin{bmatrix} Y_{01}/l_1 \\ Y_{02}/l_2 \\ Y_{03}/l_3 \\ Y_{04}/l_4 \\ Y_{05}/l_5 \\ Y_{06}/l_6 \\ Y_{07}/l_7 \end{bmatrix} = \frac{1}{12}\begin{bmatrix} -2 & 0 & 0 \\ 0 & 0 & -2 \\ 0 & 0 & 2 \\ 0 & 0 & 1 \\ 1 & 0 & -1 \\ 1 & 4 & 1 \\ 0 & 0 & 0 \end{bmatrix} \begin{bmatrix} 20 \\ 10 \\ 10 \end{bmatrix}$$

that is, $$\mathbf{Y}_0 = \mathbf{T}_0\mathbf{R}$$

Similarly,

$$\begin{bmatrix} Y_1/l_1 \\ Y_2/l_2 \\ Y_3/l_3 \\ Y_4/l_4 \\ Y_5/l_5 \\ Y_6/l_6 \\ Y_7/l_7 \end{bmatrix} = \frac{1}{l_7}\begin{bmatrix} 0 \\ 0 \\ -1 \\ -1 \\ 1 \\ -1 \\ 1 \end{bmatrix} [X]$$

that is, $$\mathbf{Y} = \mathbf{TX}$$

The flexibility matrix of the unassembled members is

$$\mathbf{F}_y = \frac{1}{AE}\begin{bmatrix} l_1 & & & & & & \\ & l_2 & & & & & \\ & & l_3 & & & & \\ & & & l_4 & & & \\ & & & & l_5 & & \\ & & & & & l_6 & \\ & & & & & & l_7 \end{bmatrix}$$

Thus,

$$\begin{aligned} \mathbf{F}_x &= \mathbf{T}^T\mathbf{F}_y\mathbf{T} \\ &= \frac{1}{AEl_7^{\,2}}(l_3^{\,3} + l_4^{\,3} + l_5^{\,3} + l_6^{\,3} + l_7^{\,3}) \end{aligned}$$

$$\mathbf{U}_x = (\mathbf{T}^T\mathbf{F}_y\mathbf{T}_0)\mathbf{R}$$

$$= \frac{1}{12l_7AE}(-20l_3{}^3 - 10l_4{}^3 + 10l_5{}^3 - 70l_6{}^3)$$

$$\mathbf{X} = -\mathbf{F}_x{}^{-1}\mathbf{U}_x = 0{\cdot}246l_7 = 1{\cdot}43 \text{ kN}$$

Thus, the member forces are

$$\mathbf{S} = \mathbf{T}_0\mathbf{R} + \mathbf{TX}$$

$$= \frac{1}{12}\begin{bmatrix} -40l_1 \\ -20l_2 \\ 20l_3 \\ 10l_4 \\ 10l_5 \\ 70l_6 \\ 0 \end{bmatrix} + 0{\cdot}246\begin{bmatrix} 0 \\ 0 \\ -l_3 \\ -l_4 \\ l_5 \\ -l_6 \\ l_7 \end{bmatrix} = \begin{bmatrix} -23{\cdot}8 \\ -11{\cdot}9 \\ 4{\cdot}26 \\ 2{\cdot}94 \\ 6{\cdot}30 \\ 16{\cdot}76 \\ 1{\cdot}43 \end{bmatrix} \text{kN}$$

The deflections of the loaded points may be obtained from

$$\mathbf{r} = \mathbf{T}_0{}^T\mathbf{F}_y(\mathbf{T}_0\mathbf{R} + \mathbf{TX})$$

3.7. Continuous Beams—Band Matrix Method

For the continuous beam, shown in Fig. 3.11, having $(n + 1)$ spans, the compatibility conditions, from which the values of the redundants $x_{1\ldots n}$ can be determined, are given by eq. (3.29):

$$\begin{bmatrix} f_{11} & f_{12} & & & \\ f_{21} & f_{22} & f_{23} & & \\ & f_{32} & f_{33} & \ddots & \\ & & \ddots & \ddots & f_{n-1,n} \\ & & & f_{n,n-1} & f_{nn} \end{bmatrix}\begin{bmatrix} x_1 \\ x_2 \\ \vdots \\ x_{n-1} \\ x_n \end{bmatrix} + \begin{bmatrix} u_1 \\ u_2 \\ \vdots \\ u_{n-1} \\ u_n \end{bmatrix} = \begin{bmatrix} 0 \\ 0 \\ \vdots \\ 0 \\ 0 \end{bmatrix} \quad (3.29)$$

or

$$\mathbf{FX} + \mathbf{U} = \mathbf{0}$$

where

$$f_{ij} = \int \frac{m_i m_j}{EI}\,\mathrm{d}s$$

$$u_i = \int \frac{m_i m_0}{EI}\,\mathrm{d}s \quad (3.30)$$

The flexibility matrix in eq. (3.29) has a banded form with all the non-zero coefficients on, or adjacent to, the leading diagonal. Such a matrix is termed 'tri-diagonal' and its form is a consequence of the limited overlap of the m_i diagrams, as shown in Fig. 3.11. An algorithm for the solution of equations

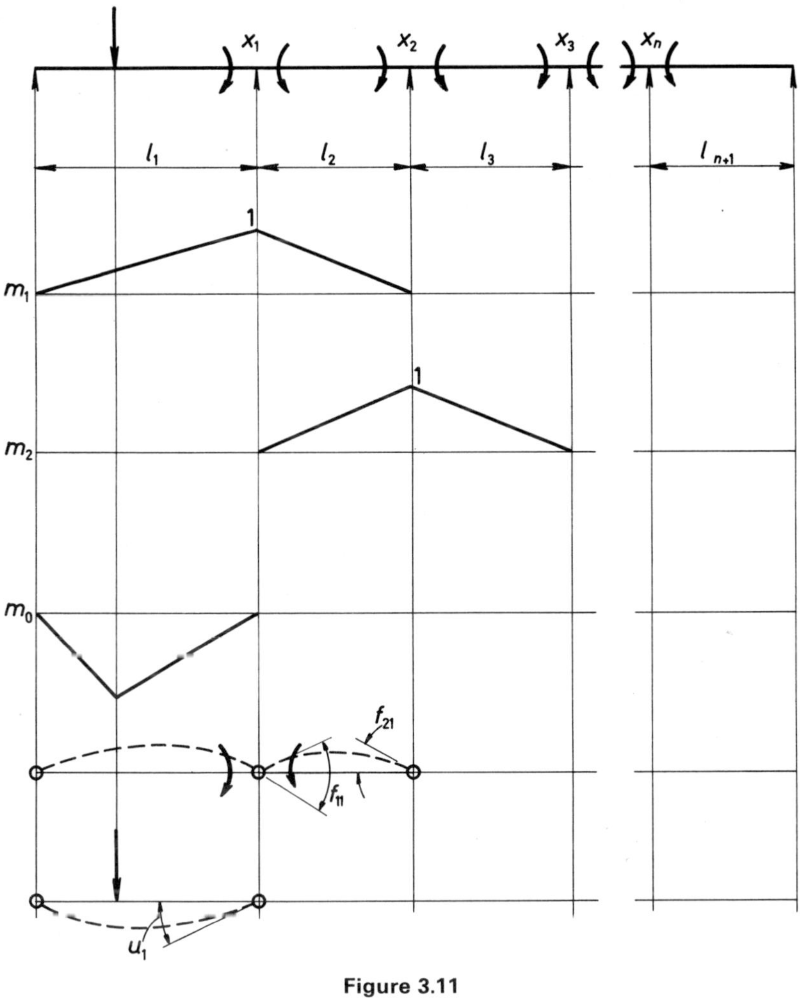

Figure 3.11

of type (3.29) will be developed in chapter 7, section 7.4. If a number of alternative loading systems have to be considered, or if influence line ordinates of the redundants are required, there will be a corresponding number of m_0 diagrams and the matrices **X** and **U** in eq. (3.29) will contain corresponding numbers of columns. The solution process, eq. (7.13), can of course be applied systematically to the **U** vectors one by one, but in general it will be more economical to compute the inverse flexibility matrix $\mathbf{F}^{-1}$ and obtain the matrix of redundants **X** by the matrix multiplication,

$$\mathbf{X} = -\mathbf{F}^{-1}\mathbf{U}$$

The flexibility influence coefficients f_{ij} and the applied load terms u_i are all angular discontinuities at the positions of the redundants in the statically determinate state. End rotations of single spans under some frequently occurring applied load systems are tabulated in Appendix B. The tabulated values apply to uniform cross-section spans only. For non-prismatic members the method of section 3.1 should be used.

Example 3.8. Influence line ordinates of the redundants x_1 and x_2 in the continuous beam of Fig. 3.12, are to be obtained corresponding to the unit load

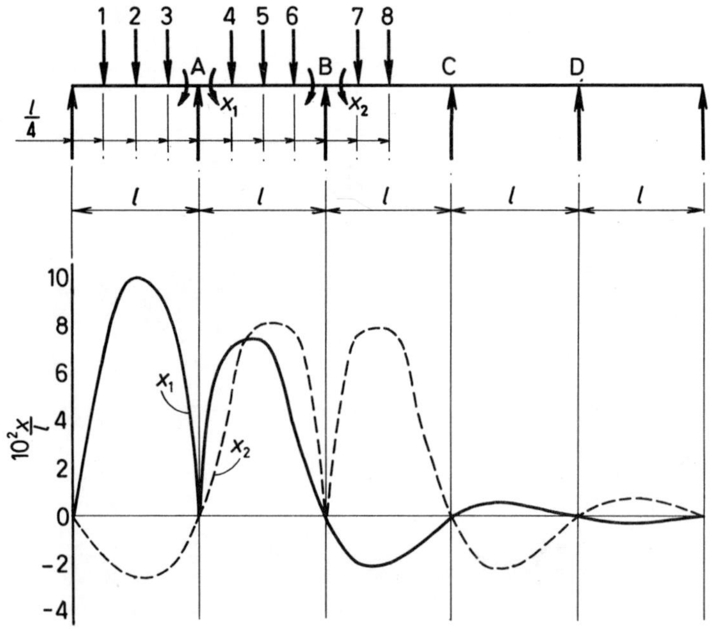

X_1	X_2
0	0
0·0628	-0·0168
0 1005	-0·0269
0·0879	-0·0236
0	0
0·0711	0·0437
0·0736	0·0807

X_1	X_2
0·0393	0·0774
0	0
-0·0191	0 0763
-0·0197	0·0790
-0·0105	0·0422
0	0
0·0052	-0·0206

X_1	X_2
0·0054	-0·0215
0·0029	-0·0117
0	0
-0·0016	0·0063
-0·0018	0·0072
-0·0011	0·0045
0	0

Figure 3.12. Example 3.8

positions 1 to 8 shown. The computation of the inverse flexibility matrix using the algorithm, eq. (7.13), is conveniently set out in tabular form:

	$\times l/6EI$											$\times 6EI/l$			
i	a_i	b_i	$\bar{a}_i$	c_{i1}	c_{i2}	c_{i3}	c_{i4}	$\bar{c}_{i1}$	$\bar{c}_{i2}$	$\bar{c}_{i3}$	$\bar{c}_{i4}$	x_{i1}	x_{i2}	x_{i3}	x_{i4}
1	4		4	1	0	0	0	1	0	0	0	0·2679			
2	4	1	3·7500	0	1	0	0	−0·25	1	0	0	−0·07177	0·2871		
3	4	1	3·7333	0	0	1	0	0·06667	−0·2667	1	0	0·01914	−0·07656	0·2871	
4	4	1	3·7321	0	0	0	1	−0·01786	0·07143	−0·2679	1	−0·004786	0·01914	−0·07178	0·2679

Thus

$$\mathbf{F}^{-1} = 6EI/l \begin{bmatrix} 0{\cdot}2679 & -0{\cdot}07177 & 0{\cdot}01914 & -0{\cdot}004785 \\ -0{\cdot}07177 & 0{\cdot}2871 & -0{\cdot}07656 & 0{\cdot}01914 \\ 0{\cdot}01914 & -0{\cdot}07656 & 0{\cdot}2871 & -0{\cdot}07177 \\ -0{\cdot}004785 & 0{\cdot}01914 & -0{\cdot}07177 & 0{\cdot}2679 \end{bmatrix}$$

Taking unit load positions, numbered 1 to 8 in Fig. 3.12, the beam rotations at supports, A, B, and C are tabulated below:

	$\times l^2/6EI$		
Unit load position	θ_A	θ_B	θ_C
1	0·2344	0	0
2	0·3750	0	0
3	0·3281	0	0
4	0·3281	0·2344	0
5	0·3750	0·3750	0
6	0·2344	0·3281	0
7	0	0·3281	0·2344
8	0	0·3750	0·3750

Due to symmetry, only one-half of the structure need be covered with the unit load positions; thus:

$$\mathbf{U} = -l^2/6EI \times$$

$$\begin{bmatrix} 0{\cdot}2344 & 0{\cdot}3750 & 0{\cdot}3281 & 0{\cdot}3281 & 0{\cdot}3750 & 0{\cdot}2344 & 0 & 0 \\ 0 & 0 & 0 & 0{\cdot}2344 & 0{\cdot}3750 & 0{\cdot}3281 & 0{\cdot}3281 & 0{\cdot}3750 \\ 0 & 0 & 0 & 0 & 0 & 0 & 0{\cdot}2344 & 0{\cdot}3750 \\ 0 & 0 & 0 & 0 & 0 & 0 & 0 & 0 \end{bmatrix}$$

and

$$\mathbf{X} = -\mathbf{F}^{-1}\mathbf{U}$$

$$= l \begin{bmatrix} 0{\cdot}06280 & 0{\cdot}1005 & 0{\cdot}08792 & 0{\cdot}07110 & 0{\cdot}07357 & 0{\cdot}03925 & -0{\cdot}01906 & -0{\cdot}01974 \\ -0{\cdot}01682 & -0{\cdot}02691 & -0{\cdot}02355 & 0{\cdot}04374 & 0{\cdot}08074 & 0{\cdot}07738 & 0{\cdot}07626 & 0{\cdot}07895 \\ 0{\cdot}00449 & 0{\cdot}00718 & 0{\cdot}00628 & -0{\cdot}01166 & -0{\cdot}02153 & -0{\cdot}02063 & 0{\cdot}04217 & 0{\cdot}07895 \\ -0{\cdot}00112 & -0{\cdot}00179 & -0{\cdot}00157 & 0{\cdot}00292 & 0{\cdot}00538 & 0{\cdot}00516 & -0{\cdot}01054 & -0{\cdot}01974 \end{bmatrix}$$

The resulting influence lines for x_1 and x_2 are plotted in Fig. 3.12. Symmetry has been used to obtain the ordinates in the right-hand half of the diagram.

PROBLEMS

3.1. Obtain the flexibility matrix for the beam shown in Fig. 3.13.

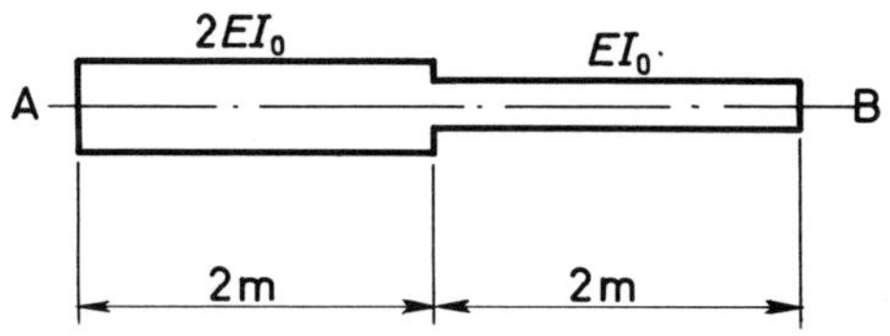

Figure 3.13. Problem 3.1

3.2. Obtain the flexibility matrix for the beam shown in Fig. 3.14. The cross-section of the beam is rectangular, of width b, and the depth varies linearly from d_0 at end A to $2d_0$ at end B.

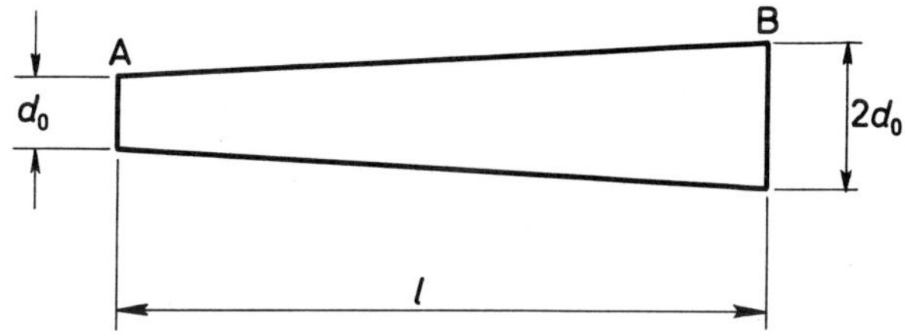

Figure 3.14. Problem 3.2

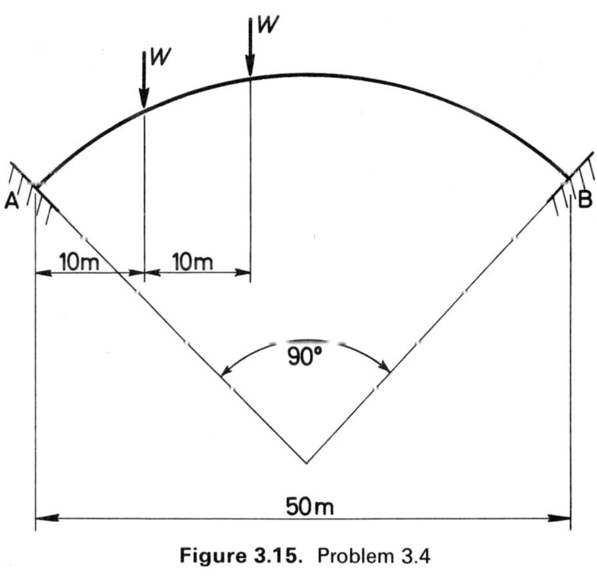

Figure 3.15. Problem 3.4

3.3. Deduce expressions, including the effects of shear, for the flexibilities of a rectangular section beam of width b, depth d, and length l, given that $E/G = 2$ and $d = l/2$.

3.4. Calculate the fixing moments at A and B and the horizontal thrust for the encastré circular segmental arch shown in Fig. 3.15.

3.5. The beam ABCD, shown in Fig. 3.16, is rigidly built-in at A and supported on a spherical bearing at D. The angle $A\hat{B}D$ is 90°. Calculate the reactions at A and D due to the load W. $EI/GJ = 5/4$.

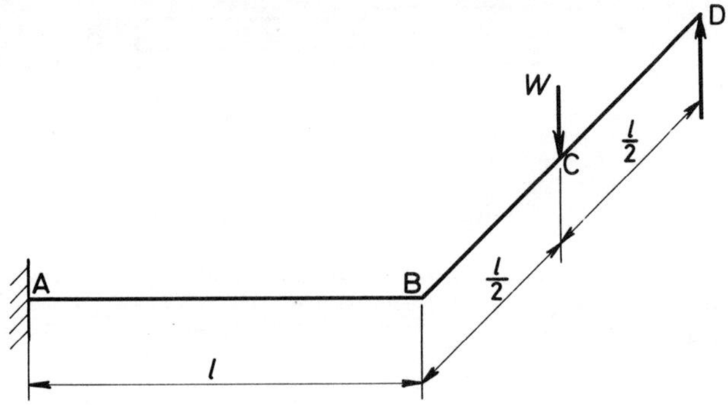

Figure 3.16. Problem 3.5

3.6. Figure 3.17 shows a plan view of a grillage comprised of three longitudinal girders AA, BB, AA and two cross-girders CC. The connections are such that only vertical forces can be transmitted and the grillage is simply-supported at A, B, A. The EI value of the longitudinals is four times thaι of the cross girders. Calculate the

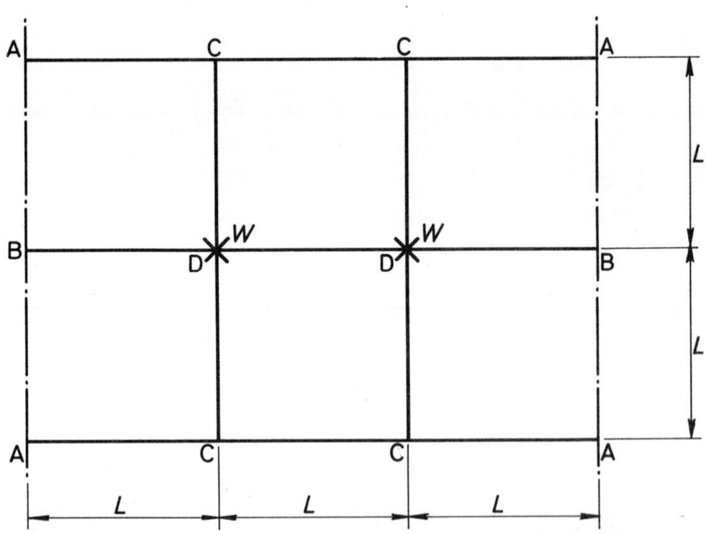

Figure 3.17. Problem 3.6

bending moment at D in the longitudinal girder BB due to the action of vertical loads W at the points D.

3.7. The girder shown in Fig. 3.18 is continuous over four spans and has a varying flexural rigidity as shown in the table. Calculate influence line ordinates for M_B, M_C, and M_D corresponding to the eight unit load positions shown.

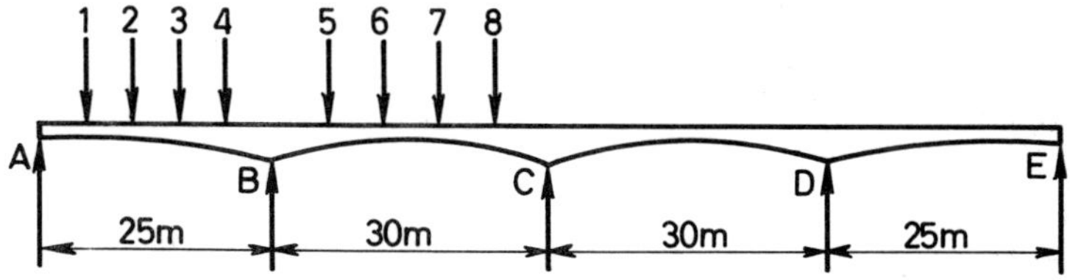

Span AB(&ED)	200(A)	206	225	259	312	391	503	662	882	1186	1600(B)
Span BC(&CD)	1600(B)	882	503	312	225	200	225	312	503	882	1600(C)

EI values (kN mm² x 10^{-9}) at $\frac{1}{10}$th span points

Figure 3.18. Problem 3.7

(The computations may follow the scheme of section 3.7, or the program of section 8.2 may be used.)

4. The Stiffness Method

The basic principles of the stiffness method were outlined in chapter 1, section 1.7. In this chapter the method will be developed in more detail with some emphasis on a formulation suitable for digital computers. The stiffness matrix of a beam element has already been quoted, eq. (1.27), and this matrix contains all the information needed to represent the behaviour of the beam within a structure. Clearly, other types of structural element are possible, and further examples will be considered in chapter 6.

The general form of the force-displacement relationships for the ith element in a structure is

$$\mathbf{S}_i = \mathbf{k}_i \mathbf{s}_i \tag{4.1}$$

in which there is a strict correspondence, in position and direction, between the individual nodal forces in $\mathbf{S}_i$ and the nodal displacements in $\mathbf{s}_i$. For the

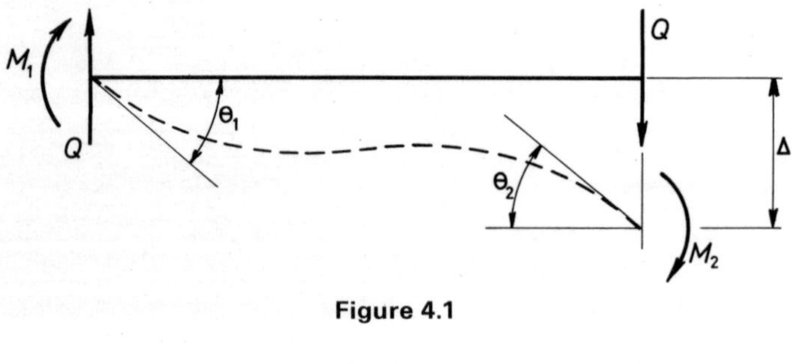

Figure 4.1

uniform beam shown in Fig. 4.1, eq. (4.1) becomes

$$\begin{bmatrix} M_1 \\ M_2 \\ Q \end{bmatrix} = EI \begin{bmatrix} 4/l & 2/l & -6/l^2 \\ 2/l & 4/l & -6/l^2 \\ -6/l^2 & -6/l^2 & 12/l^3 \end{bmatrix} \begin{bmatrix} \theta_1 \\ \theta_2 \\ \Delta \end{bmatrix} \tag{4.2}$$

It will be recognized that the stiffness matrix in eq. (4.2) is obtained from that in eq. (1.27) by the elimination of the second row and column, thus removing a rigid body translation of the beam.

In chapter 1, the assembly of a stiffness matrix of a structure was achieved by a process of selecting appropriate stiffnesses from elements and combining these to form equilibrium equations corresponding to the nodal degrees of freedom. We shall now consider how this process of assembly of the structure stiffness matrix from that of its elements may be performed automatically.

4.1. Stiffness Transformation

Equation (4.1) represents the force-displacement relationships for the ith element in a structure. Suppose there are n elements in the structure, then the force-displacement relationships for the complete, though unconnected, structure may be represented as follows:

$$\begin{bmatrix} S_1 \\ S_2 \\ \vdots \\ S_n \end{bmatrix} = \begin{bmatrix} k_1 & & & \\ & k_2 & & \\ & & \ddots & \\ & & & k_n \end{bmatrix} \begin{bmatrix} s_1 \\ s_2 \\ \vdots \\ s_n \end{bmatrix}$$

that is, $$\mathbf{S} = \mathbf{k}\mathbf{s} \tag{4.3}$$

The equilibrium conditions for the complete structure, relating applied loads and internal nodal displacements, are, eq. (1.30),

$$\mathbf{K}\mathbf{r} = \mathbf{R} \tag{4.4}$$

If the two sets of displacements $\mathbf{r}$ and $\mathbf{s}$ are connected by the transformation

$$\mathbf{s} = \mathbf{A}\mathbf{r} \tag{4.5}$$

then, from the principle of contragredience, eqns. (1.35) and (1.39),

$$\mathbf{R} = \mathbf{A}^T\mathbf{S} \tag{4.6}$$

$$= \mathbf{A}^T\mathbf{k}\mathbf{s}, \text{ substituting for } \mathbf{S} \text{ from (4.3)}$$

$$= \mathbf{A}^T\mathbf{k}\mathbf{A}\mathbf{r}, \text{ substituting for } \mathbf{s} \text{ from (4.5)} \tag{4.7}$$

Comparing eqns. (4.7) and (4.4) it is evident that

$$\mathbf{K} = \mathbf{A}^T\mathbf{k}\mathbf{A} \tag{4.8}$$

The structure stiffness matrix $\mathbf{K}$ can thus be determined from the unassembled element stiffnesses $\mathbf{k}$ by the transformation $\mathbf{A}^T\mathbf{k}\mathbf{A}$.

Example 4.1. The structure, shown in Fig. 4.2a, consists of a fixed-end beam AB supported by cables EC and DF. The cables lie in the vertical plane and have an extensional rigidity $EA = EI/20$ (metre units), where EI is the flexural rigidity of the beam. The loading consists of a uniformly distributed vertical load w over the middle third of the span.

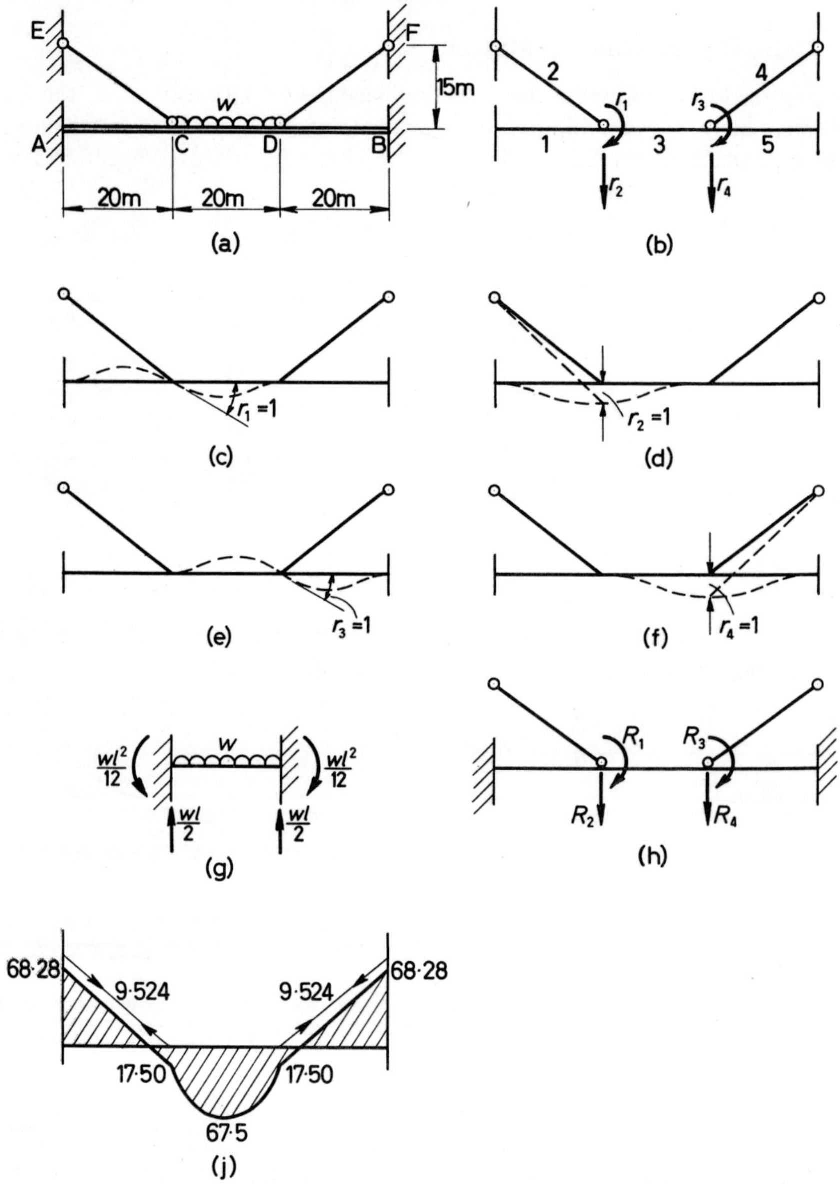

Figure 4.2. Example 4.1

The structure is divided into five elements, three beam elements, numbered 1, 3, and 5 in Fig. 4.2b, and the cable elements 2 and 4. We shall ignore symmetry for the time being and investigate later how this may be taken into account. There are four degrees of freedom in the structure, taken as $r_{1,2,3,4}$ in Figs. 4.2c, d, e, and f.

The stiffness matrix of each beam element may be obtained from eq. (4.2), and for each cable element the stiffness is simply EA/l being the axial force required to produce unit axial extension of the cable.

Thus, the stiffness matrix of the unconnected elements is

$$\mathbf{k} = \frac{EI}{10^3}\left[\begin{array}{ccc|c|ccc|c|ccc} 200 & 100 & -15 & & & & & & & & \\ 100 & 200 & -15 & & & & & & & & \\ -15 & -15 & 1{\cdot}5 & & & & & & & & \\ \hline & & & 2 & & & & & & & \\ \hline & & & & 200 & 100 & -15 & & & & \\ & & & & 100 & 200 & -15 & & & & \\ & & & & -15 & -15 & 1{\cdot}5 & & & & \\ \hline & & & & & & & 2 & & & \\ \hline & & & & & & & & 200 & 100 & -15 \\ & & & & & & & & 100 & 200 & -15 \\ & & & & & & & & 15 & 15 & 1{\cdot}5 \end{array}\right] \qquad (4.9)$$

The displacement transformation matrix $\mathbf{A}$ is obtained by considering the unit displacement diagrams, Figs. 4.2c, d, e, and f, and writing down, in columns, the corresponding individual element nodal displacements. The elements are taken in the same order as that used in assembling the matrix $\mathbf{k}$.

Thus,

$$\mathbf{A} = \begin{array}{c} \\ \end{array}\begin{array}{|c|c|c|c|l} \multicolumn{1}{c}{r_1 = 1} & \multicolumn{1}{c}{r_2 = 1} & \multicolumn{1}{c}{r_3 = 1} & \multicolumn{1}{c}{r_4 = 1} & \\ \hline 0 & 0 & 0 & 0 & \\ \hline 1 & 0 & 0 & 0 & \text{element 1} \\ \hline 0 & 1 & 0 & 0 & \\ \hline 0 & \frac{3}{5} & 0 & 0 & \text{element 2} \\ \hline 1 & 0 & 0 & 0 & \\ \hline 0 & 0 & 1 & 0 & \text{element 3} \\ \hline 0 & -1 & 0 & 1 & \\ \hline 0 & 0 & 0 & \frac{3}{5} & \text{element 4} \\ \hline 0 & 0 & 1 & 0 & \\ \hline 0 & 0 & 0 & 0 & \text{element 5} \\ \hline 0 & 0 & 0 & -1 & \\ \hline \end{array} \qquad (4.10)$$

In computing the triple product $\mathbf{A}^T\mathbf{kA}$ it is convenient to form $\mathbf{kA}$ first as this is useful later in computing element forces.

$$\mathbf{kA} = \frac{EI}{10^3}\begin{array}{|c|c|c|c|} \hline 100 & -15 & 0 & 0 \\ 200 & -15 & 0 & 0 \\ -15 & 1{\cdot}5 & 0 & 0 \\ \hline 0 & 1{\cdot}2 & 0 & 0 \\ \hline 200 & 15 & 100 & -15 \\ 100 & 15 & 200 & -15 \\ -15 & -1{\cdot}5 & -15 & 1{\cdot}5 \\ \hline 0 & 0 & 0 & 1{\cdot}2 \\ \hline 0 & 0 & 200 & 15 \\ 0 & 0 & 100 & 15 \\ 0 & 0 & -15 & -1{\cdot}5 \\ \hline \end{array} \qquad (4.11)$$

$$\mathbf{K} = \mathbf{A}^T\mathbf{k}\mathbf{A} = \frac{EI}{10^3}\begin{bmatrix} 400 & 0 & 100 & -15 \\ 0 & 3{\cdot}72 & 15 & -1{\cdot}5 \\ 100 & 15 & 400 & 0 \\ -15 & -1{\cdot}5 & 0 & 3{\cdot}72 \end{bmatrix} \tag{4.12}$$

It should be noted that the stiffness matrix is symmetric, as anticipated in section 1.7, eq. (1.28). The flexibility matrix **F** is the inverse of the stiffness matrix, and is

$$\mathbf{F} = \mathbf{K}^{-1} = \frac{1}{EI}\begin{bmatrix} 3{\cdot}61 & 13{\cdot}86 & -1{\cdot}42 & 20{\cdot}15 \\ 13{\cdot}86 & 445{\cdot}0 & -20{\cdot}15 & 235{\cdot}3 \\ -1{\cdot}42 & -20{\cdot}15 & 3{\cdot}61 & -13{\cdot}86 \\ 20{\cdot}15 & 235{\cdot}3 & -13{\cdot}86 & 445{\cdot}0 \end{bmatrix} \tag{4.13}$$

4.2. Applied Loads

The application of the external loads to the primary, constrained, structure results in a set of element nodal forces $\mathbf{S}_0$ throughout the structure and a set of constraining forces at the nodes. The constraining forces, reversed in sign, constitute the loading **R** on the structure for determining the complementary solution.

Thus, in example 4.1, Figs. 4.2g and h:

$$\mathbf{S}_0 = \begin{bmatrix} 0 \\ 0 \\ 0 \\ 0 \\ -wl^2/12 \\ wl^2/12 \\ wl/2 \\ 0 \\ 0 \\ 0 \\ 0 \end{bmatrix} \begin{matrix} \left.\begin{matrix} \\ \\ \\ \end{matrix}\right\} 1 \\ \} 2 \\ \left.\begin{matrix} \\ \\ \\ \end{matrix}\right\} 3 \\ \} 4 \\ \left.\begin{matrix} \\ \\ \\ \end{matrix}\right\} 5 \end{matrix} = w\begin{bmatrix} 0 \\ 0 \\ 0 \\ 0 \\ -33{\cdot}33 \\ 33{\cdot}33 \\ 10 \\ 0 \\ 0 \\ 0 \\ 0 \end{bmatrix} \tag{4.14}$$

$$\mathbf{R} = w \begin{bmatrix} 33{\cdot}33 \\ 10 \\ -33{\cdot}33 \\ 10 \end{bmatrix} \tag{4.15}$$

The nodal displacements caused by **R** can now be calculated:

$$\mathbf{r} = \mathbf{K}^{-1}\mathbf{R}$$

$$= \frac{10^3 w}{EI} \begin{bmatrix} 0{\cdot}5078 \\ 7{\cdot}937 \\ -0{\cdot}5078 \\ 7{\cdot}937 \end{bmatrix} \tag{4.16}$$

4.3. Element Nodal Forces

The element nodal forces corresponding to loading **R** may be obtained as follows, from eq. (4.3), substituting **s** from eq. (4.5),

$$\mathbf{S} = \mathbf{kAr} \tag{4.17}$$

In example 4.1

$$\mathbf{S} = w \left[\begin{array}{r} -68{\cdot}28 \\ -17{\cdot}50 \\ 4{\cdot}289 \\ \hline 9{\cdot}524 \\ \hline 50{\cdot}78 \\ -50{\cdot}78 \\ 0 \\ \hline 9{\cdot}524 \\ \hline 17{\cdot}50 \\ 68{\cdot}28 \\ -4{\cdot}289 \end{array} \right] \tag{4.18}$$

and the total member forces are obtained by addition of the particular and complementary solutions:

$$\mathbf{S} + \mathbf{S}_0 = w \left[\begin{array}{r} -68{\cdot}28 \\ -17{\cdot}50 \\ 4{\cdot}289 \\ \hline 9{\cdot}524 \\ \hline 17{\cdot}45 \\ -17{\cdot}45 \\ 10{\cdot}00 \\ \hline 9{\cdot}524 \\ \hline 17{\cdot}50 \\ 68{\cdot}28 \\ -4{\cdot}289 \end{array}\right] \quad \text{(kN and metre units)} \tag{4.19}$$

The bending moment diagram for the beam is shown in Fig. 4.2j.

4.4. Treatment of Symmetry

The computational work in example 4.1 could have been considerably reduced by taking advantage of the structural symmetry. In general, two cases arise:

(a) when the structure is symmetrical but the applied load is not, and

(b) when both the structure and the applied load are symmetrical.

Case (a) can be handled by resolving nodal displacements and applied loads into symmetrical and antisymmetrical components. The procedure is conveniently illustrated by application to example 4.1. Ignoring the symmetry of the actual loading for the time being, the symmetrical and antisymmetrical displacement components $q_{1,2,3,4}$ are shown in Figs. 4.3a, b respectively.

The relationships between the displacements $\mathbf{r}$, Fig. 4.2, and $\mathbf{q}$, Fig. 4.3, are as follows:

$$r_1 = q_1 + q_3$$

$$r_2 = q_2 + q_4$$

$$r_3 = -q_1 + q_3$$

$$r_4 = q_2 - q_4$$

that is,

$$\mathbf{r} = \mathbf{B}\mathbf{q} \tag{4.20}$$

where

$$\mathbf{B} = \begin{bmatrix} 1 & 0 & 1 & 0 \\ 0 & 1 & 0 & 1 \\ -1 & 0 & 1 & 0 \\ 0 & 1 & 0 & -1 \end{bmatrix} \tag{4.21}$$

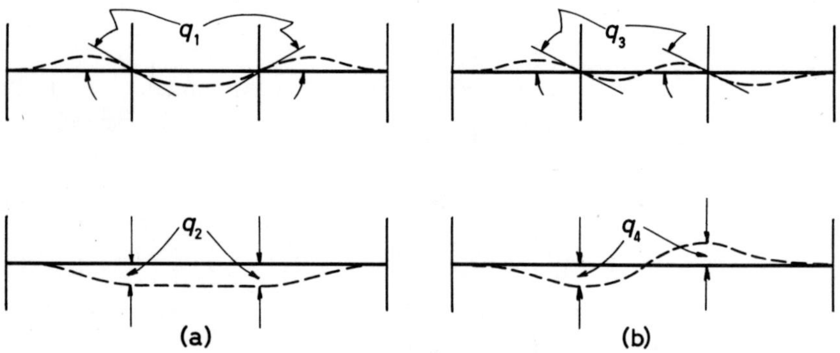

Figure 4.3. (a) Symmetrical displacement components; (b) anti-symmetrical displacement components

The transformation from the r system to the q system is, from eq. (4.8),

$$\mathbf{K}_q = \mathbf{B}^T\mathbf{K}\mathbf{B} \tag{4.22}$$

in which the stiffness matrix $\mathbf{K}$ is given by eq. (4.12), thus:

$$\mathbf{K}_q = \frac{EI}{10^3}\left[\begin{array}{cc|cc} 600 & -30 & 0 & 0 \\ -30 & 4{\cdot}44 & 0 & 0 \\ \hline 0 & 0 & 1000 & 30 \\ 0 & 0 & 30 & 10{\cdot}44 \end{array}\right] \tag{4.23}$$

The 'uncoupled' nature of the submatrices corresponding to symmetric and antisymmetric displacements should be noted.

For symmetrical loading, the antisymmetrical displacements can be disregarded, thus

$$\frac{EI}{10^3}\begin{bmatrix} 600 & -30 \\ -30 & 4{\cdot}44 \end{bmatrix}\begin{bmatrix} q_1 \\ q_2 \end{bmatrix} = \begin{bmatrix} Q_1 \\ Q_2 \end{bmatrix} \tag{4.24}$$

The symmetrical and antisymmetrical components of the applied loads, $Q_{1,2,3,4}$, are, from eq. (4.6),

$$\mathbf{Q} = \mathbf{B}^T\mathbf{R}$$

$$= \begin{bmatrix} 1 & 0 & -1 & 0 \\ 0 & 1 & 0 & 1 \\ 1 & 0 & 1 & 0 \\ 0 & 1 & 0 & -1 \end{bmatrix} w \begin{bmatrix} 33{\cdot}33 \\ 10 \\ -33{\cdot}33 \\ 10 \end{bmatrix} = w \begin{bmatrix} 66{\cdot}67 \\ 20 \\ 0 \\ 0 \end{bmatrix} \quad (4.25)$$

Hence

$$\begin{bmatrix} q_1 \\ q_2 \end{bmatrix} = \frac{10^3 w}{EI} \begin{bmatrix} 0{\cdot}5078 \\ 7{\cdot}937 \end{bmatrix} \quad (4.26)$$

which agrees with the previous solution, eq. (4.16). If the applied loading is unsymmetrical the antisymmetrical loads Q_3 and Q_4 will be non-zero and the process consists of the solution of two sets of two equations. When the applied loading is symmetrical the antisymmetrical displacements and applied load components can of course be disregarded throughout.

4.5. Support Displacements, Prestrain, and Thermal Effects

In the basic equilibrium equation (4.4) the matrix **R** represents the loads applied to the nodes of the structure for determination of the complementary solution. The loads represented by $-\mathbf{R}$ consist of the constraining forces applied to the structure, in the particular solution, required to maintain the specified nodal displacements. Usually, the nodal displacements are made zero, but the particular solution can be one in which specific non-zero displacements are imposed. This device enables effects such as support displacement, prestrain, and thermal expansion to be included in the analysis.

Example 4.2. In the continuous beam shown in Fig. 4.4, the support B is considered to be subjected to a vertical settlement δ. It is required to determine the resulting bending moment distribution in the beam.

The stiffness matrix is easily shown to be

$$\mathbf{K} = \frac{EI}{l} \begin{bmatrix} 8 & 2 \\ 2 & 8 \end{bmatrix} \quad (4.27)$$

and the inverse

$$\mathbf{K}^{-1} = \frac{l}{EI} \begin{bmatrix} 2/15 & -1/30 \\ -1/30 & 2/15 \end{bmatrix} \quad (4.28)$$

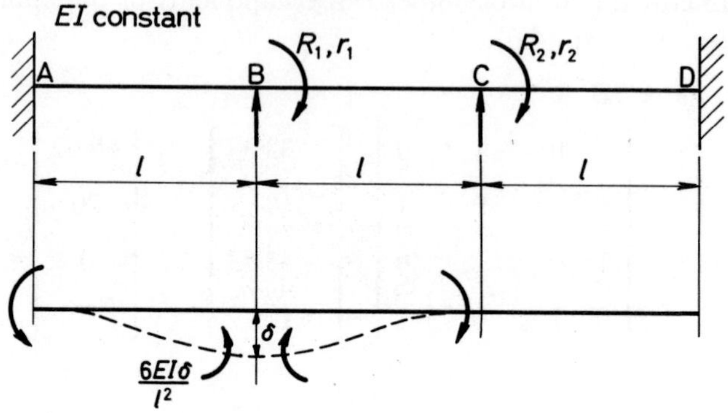

Figure 4.4. Example 4.2

From Fig. 4.4,

$$\mathbf{R} = \begin{bmatrix} 0 \\ -\dfrac{6EI\delta}{l^2} \end{bmatrix} \tag{4.29}$$

Thus

$$\mathbf{r} = \mathbf{K}^{-1}\mathbf{R} = \begin{bmatrix} \delta/5l \\ -4\delta/5l \end{bmatrix} \tag{4.30}$$

The bending moments in the spans may be calculated using eq. (4.2), for example in span AB,

$$M_A = -28EI\delta/5l^2$$

$$M_B = -26EI\delta/5l^2$$

4.6. Continuous Beams

We have seen that the stiffness matrix of a structure can be assembled automatically by carrying out the triple matrix multiplication of eq. (4.8). The matrix **k** contains the stiffness matrices of the individual disconnected elements, and the matrix **A** relates the element nodal displacements to the structure nodal displacements. The transformation matrix **A** is entirely dependent on the geometry of the structure and, in cases where the structural geometry is particularly easily described, the transformation method can be dispensed with and the structure stiffness matrix written down by inspection. The continuous beam and the plane rigid frame, dealt with in section 4.7, are examples of this type of structure.

A general n span continuous beam is shown in Fig. 4.5. The supports are taken to have given stiffnesses in the vertical direction, $\beta_1, \beta_2, \ldots, \beta_{n+1}$.

There are two displacements to be considered at each support, and these are numbered from left to right as shown in Fig. 4.5. In writing down the stiffness matrix it is helpful in the first instance to sketch diagrams of the

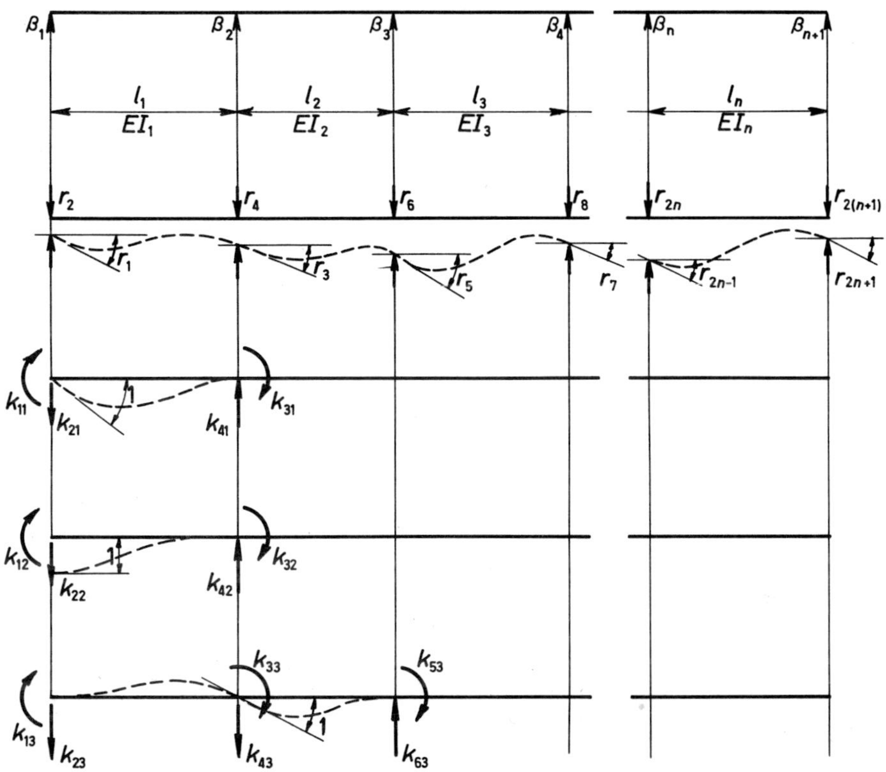

Figure 4.5. General, *n* span, continuous beam

deflected form of the structure, corresponding to unit values of each displacement in turn. It should be remembered that, in doing this, all the nodal displacements are made zero except the one being given unit value. Only the first few diagrams need be sketched since the pattern which develops in the stiffness matrix will soon become apparent and the remainder of the matrix can then be written in.

Taking $$\frac{EI_i}{l_i} = \alpha_i$$

Then, referring to the beam element stiffness matrix eq. (4.2),

$$\left.\begin{aligned}
k_{11} &= 4\alpha_1 \\
k_{12} &= k_{21} = \frac{6\alpha_1}{l_1} \\
k_{13} &= k_{31} = 2\alpha_1 \\
k_{14} &= k_{41} = -\frac{6\alpha_1}{l_1} \\
k_{22} &= \beta_1 + \frac{12\alpha_1}{l_1^2} \\
k_{23} &= k_{32} = \frac{6\alpha_1}{l_1} \\
k_{24} &= k_{42} = -\frac{12\alpha_1}{l_1^2} \\
k_{33} &= 4\alpha_1 + 4\alpha_2 \\
k_{34} &= k_{43} = \frac{6\alpha_2}{l_2} - \frac{6\alpha_1}{l_1} \\
k_{35} &= k_{53} = 2\alpha_2 \\
k_{36} &= k_{63} = -\frac{6\alpha_2}{l_2}
\end{aligned}\right\} \qquad (4.31)$$

Thus the stiffness matrix can be written down as in eq. (4.32) which contains the lower half of the symmetric matrix.

$$
\mathbf{K} = \begin{array}{c|cccccccccc|}
 & & & & & & & & & & \\
1 & 4\alpha_1 & & & & & & & & & \\
2 & \frac{6\alpha_1}{l_1} & \beta_1 + \frac{12\alpha_1}{l_1^2} & & & & & & & \text{Symmetric} & \\
3 & 2\alpha_1 & \frac{5\alpha_1}{l_1} & 4(\alpha_1 + \alpha_2) & & & & & & & \\
4 & -\frac{6\alpha_1}{l_1} & -\frac{12\alpha_1}{l_1^2} & 6\left(\frac{\alpha_2}{l_2} - \frac{\alpha_1}{l_1}\right) & \beta_2 + 12\left(\frac{\alpha_1}{l_1^2} + \frac{\alpha_2}{l_2^2}\right) & & & & & & \\
5 & & & 2\alpha_2 & \frac{6\alpha_2}{l_2} & 4(\alpha_2 + \alpha_3) & & & & & \\
6 & & & -\frac{6\alpha_2}{l_2} & -\frac{12\alpha_2}{l_2^2} & 6\left(\frac{\alpha_3}{l_3} - \frac{\alpha_2}{l_2}\right) & \beta_3 + 12\left(\frac{\alpha_2}{l_2^2} + \frac{\alpha_3}{l_3^2}\right) & & & & \\
 & & & & & \ddots & \ddots & \ddots & & & \\
2n+1 & & & & & & 2\alpha_n & \frac{6\alpha_n}{l_n} & 4\alpha_n & & \\
2(n+1) & & & & & & -\frac{6\alpha_n}{l_n} & -\frac{12\alpha_n}{l_n^2} & -\frac{6\alpha_n}{l_n} & \beta_{n+1} + \frac{12\alpha_n}{l_n^2} & \\
 & & & & & & 2n-1 & 2n & 2n+1 & 2(n+1) &
\end{array}
\qquad (4.32)
$$

The continuous beam stiffness matrix takes a somewhat simpler form if the supports are rigid, i.e., $\beta_i = \infty$, $1 \leqslant i \leqslant n + 1$. In this case, the matrix can be obtained simply by deletion of the even-numbered rows and columns in matrix (4.32). Thus

$$
\mathbf{K} = \begin{array}{r}1\\2\\3\\4\\ \\ \\n\\n+1\end{array}
\left[\begin{array}{cccccccc}
4\alpha_1 & & & & & & & \\
2\alpha_1 & 4(\alpha_1+\alpha_2) & & & & \text{Symmetric} & & \\
 & 2\alpha_2 & 4(\alpha_2+\alpha_3) & & & & & \\
 & & 2\alpha_3 & 4(\alpha_3+\alpha_4) & & & & \\
 & & & \ddots & \ddots & & & \\
 & & & & 2\alpha_{n-1} & 4(\alpha_{n-1}+\alpha_n) & & \\
 & & & & & 2\alpha_n & 4\alpha_n &
\end{array}\right]
\begin{array}{c} \\ \\ \\ \\ \\ \\ \\ \\ \end{array}
\qquad (4.33)
$$

$$\qquad\qquad\qquad\qquad\qquad\qquad n \qquad n+1$$

The unit displacement diagrams in Fig. 4.5 show that there is a limited spread of the cross stiffnesses in each case. Due to the nodal restraints the cross stiffnesses beyond the adjacent supports are always zero. This produces the banded form of the stiffness matrices in eqns. (4.32) and (4.33). Band matrices are very common in structural analysis using the stiffness method and special methods have been developed for their solution.[5, 7, 20, 22, 25] Further attention will be given to band matrix solutions in chapter 7, where it will be shown that equations of type (4.33) can be solved in a particularly simple fashion.

4.7. Plane Rigid Frames

This is another example of a structural type for which the stiffness matrix can often be assembled by inspection using the prismatic member stiffnesses of eq. (4.2).

Example 4.3. A two-storey rigid frame in which all members have the same uniform section, is shown in Fig. 4.6. The nodal displacements $r_{1\ldots 6}$ are also shown in Fig. 4.6. The stiffness matrix **K** can be written down by inspection as in eq. (4.34).

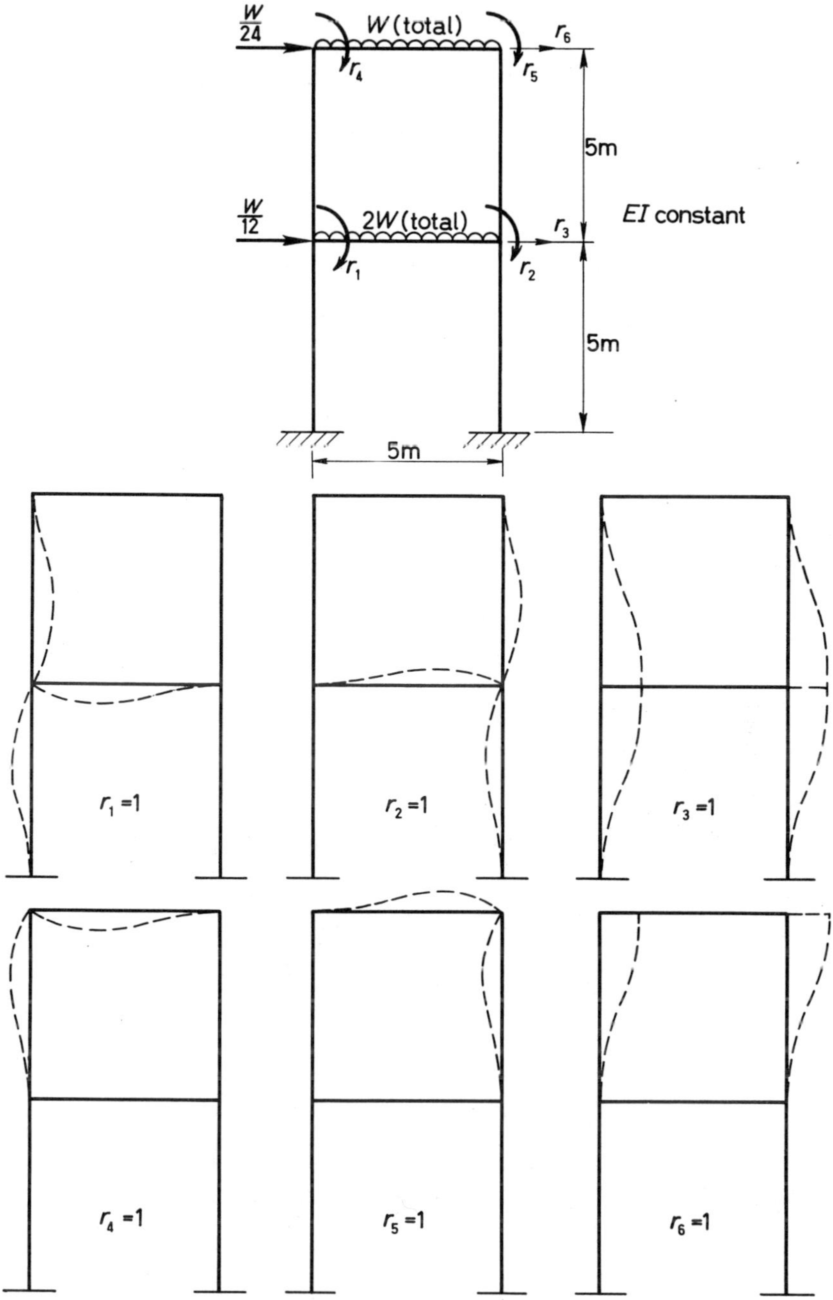

Figure 4.6. Example 4.3

$$\mathbf{K} = \frac{EI}{10^3}\begin{bmatrix} 2400 & & & & & \text{Symmetric} \\ 400 & 2400 & & & & \\ 0 & 0 & 384 & & & \\ 400 & 0 & 240 & 1600 & & \\ 0 & 400 & 240 & 400 & 1600 & \\ -240 & -240 & -192 & -240 & -240 & 192 \end{bmatrix} \tag{4.34}$$

The applied load matrix is

$$\mathbf{R} = \frac{W}{24}\begin{bmatrix} 20 \\ -20 \\ 2 \\ 10 \\ -10 \\ 1 \end{bmatrix} \tag{4.35}$$

Solution of the equilibrium equations in the form $\mathbf{r} = \mathbf{K}^{-1}\mathbf{R}$ will yield the nodal displacements $\mathbf{r}$.

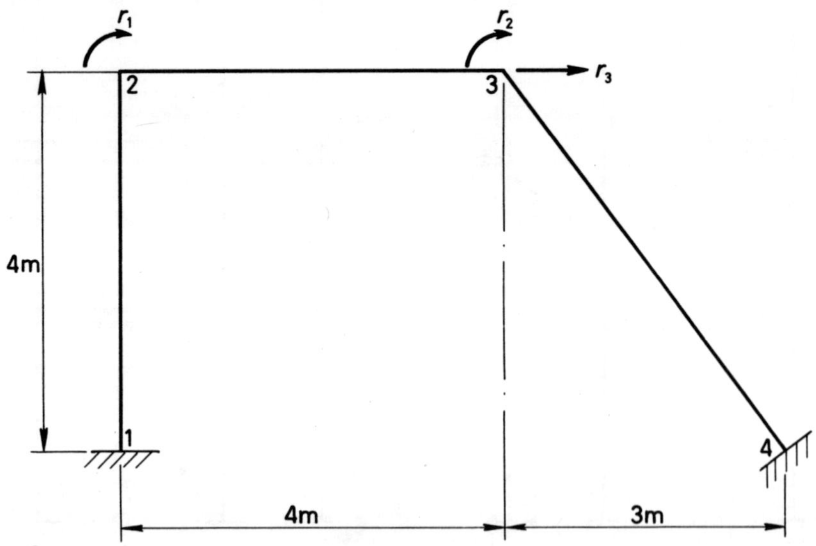

Figure 4.7. Example 4.4

Example 4.4. The structure shown in Fig. 4.7 is a rigid frame with an inclined column. If the members are considered to be inextensible, the horizontal deflections at 2 and 3 will be equal. The transformation method will be used to establish the stiffness matrix relating to displacements $r_{1,2,3}$. Taking the members in the order 1–2, 2–3, and 3–4,

$$\mathbf{k} = EI \begin{bmatrix} 1 & \frac{1}{2} & -6/4^2 & & & & & & \\ \frac{1}{2} & 1 & -6/4^2 & & & & & & \\ -6/4^2 & -6/4^2 & 12/4^3 & & & & & & \\ & & & 1 & \frac{1}{2} & -6/4^2 & & & \\ & & & \frac{1}{2} & 1 & -6/4^2 & & & \\ & & & -6/4^2 & -6/4^2 & 12/4^3 & & & \\ & & & & & & 4/5 & 2/5 & -6/5^2 \\ & & & & & & 2/5 & 4/5 & -6/5^2 \\ & & & & & & -6/5^2 & -6/5^2 & 12/5^3 \end{bmatrix} \tag{4.36}$$

From eq. (4.5), the transformation matrix **A** is

$$\mathbf{A} = \begin{bmatrix} 0 & 0 & 0 \\ 1 & 0 & 0 \\ 0 & 0 & 1 \\ 1 & 0 & 0 \\ 0 & 1 & 0 \\ 0 & 0 & -\frac{3}{4} \\ 0 & 1 & 0 \\ 0 & 0 & 0 \\ 0 & 0 & \frac{5}{4} \end{bmatrix} \tag{4.37}$$

Thus

$$\mathbf{K} = \mathbf{A}^T\mathbf{k}\mathbf{A} = \frac{EI}{20^3}\begin{bmatrix} 16\,000 & 4000 & -750 \\ 4000 & 14\,400 & -150 \\ -750 & -150 & 3544 \end{bmatrix} \tag{4.38}$$

4.8. Automatic Assembly of Stiffness Matrix—Code Number Method

The transformation method of stiffness matrix assembly, eq. 4.8, has the disadvantage of requiring a large amount of computer storage for the matrices **A** and **k**. The method of assembly by inspection is suitable only for hand calculations and a more formal and logical approach is needed for a

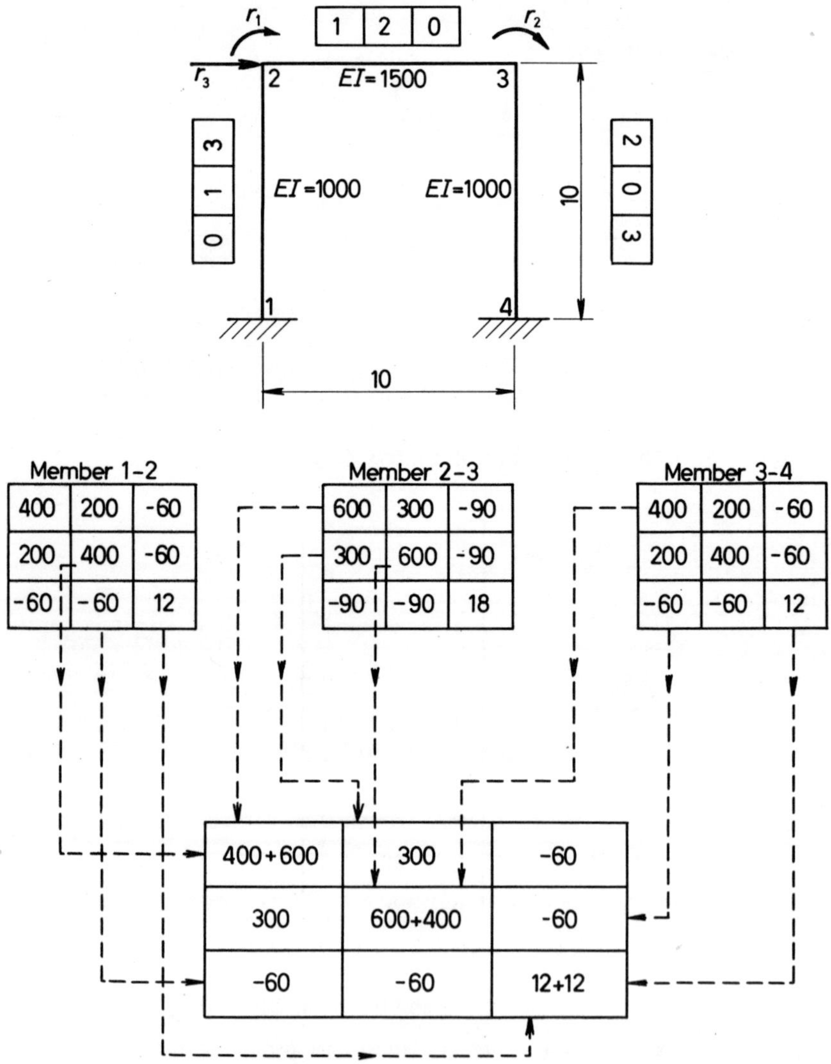

Figure 4.8. Assembly of stiffness matrix by code number method

computer method. The 'code number' method was first suggested by Tezcan,[35] and is very suitable for this purpose.

Figure 4.8 shows a portal frame along with the stiffness matrices of the three member elements. The process of assembling the structure stiffness matrix **K** is one of transferring individual element stiffnesses into appropriate positions in the matrix **K**. The 'address', or location, of a stiffness in an element matrix is defined by its row and column numbers. Similarly in the structure matrix, the address of a particular stiffness is defined by its row and column numbers. To generate the structure stiffness matrix from the element stiffnesses it is necessary to relate addresses in the element system with those in the structure system.

Considering, for example, member 1–2: the element system comprises three degrees of freedom, θ_1 and θ_2, the rotations at 1 and 2, and Δ_{1-2}, the horizontal displacement of 2 relative to 1. The addresses of stiffnesses in the element stiffness matrix are defined by the combinations of row and column numbers, 11, 12, 13; 21, 22, 23; and 31, 32, 33. If three locations are reserved for the code number for this member and the locations are named 1, 2, and 3, then the name of each location, paired with itself and the remaining two names, will form the sequence of addresses of the element stiffnesses. The address formed by combining a location name with itself, i.e., 11, 22, 33, relates of course to a direct stiffness on the diagonal of the matrix. In each location in the element code number is inserted the number of the corresponding displacement in the structure system. Thus, the code number for member 1–2 in Fig. 4.8 is

1	2	3	← addresses in element
0	1	3	← addresses in structure

the zero being inserted in the first location since this particular element displacement does not exist in the structure. Now the addresses of the element stiffnesses are defined by the names of the boxes, grouped in pairs, and the addresses of the structure stiffnesses are defined by the contents of the boxes taken in pairs. Thus, for member 1–2,

$$k_{22} \rightarrow K_{11}$$

$$k_{23} \rightarrow K_{13} \rightarrow K_{31}$$

$$k_{33} \rightarrow K_{33}$$

The other two members are dealt with in the same way and the structure stiffness matrix assembled as in Fig. 4.8. It should be noted that a double transfer is required in the case of an off-diagonal stiffness, such as k_{23}.

PROBLEMS

4.1. Assemble a stiffness matrix for the two-storey frame with rigid joints shown in Fig. 4.9. Ignore axial strains in the members and take the nodal displacements in the order 1 to 9 shown in the figure.

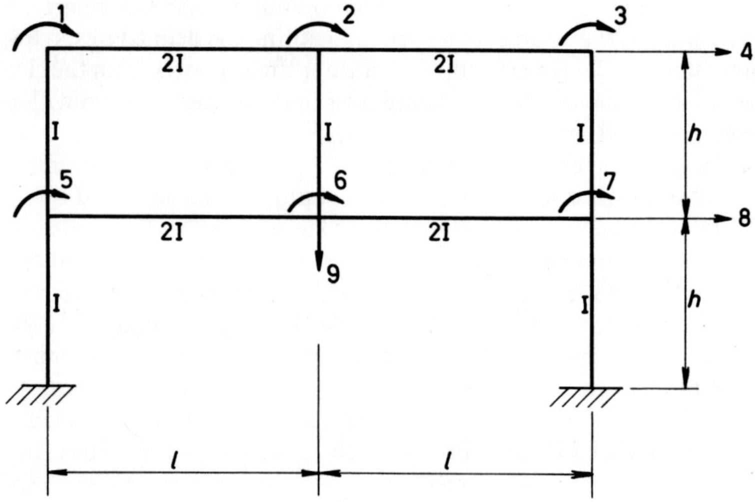

Figure 4.9. Problem 4.1

4.2. The two-span continuous beam shown in Fig. 4.10 is supported elastically at A, B, and C. Each support has a vertical stiffness of 3000 kN/m and a rotational stiffness of 6000 kNm/radian. $EI = 10\ 000$ kNm^2. Assemble a stiffness matrix in symmetrical, antisymmetrical form and calculate the vertical deflections of the supports. Draw the bending moment diagram for the beam.

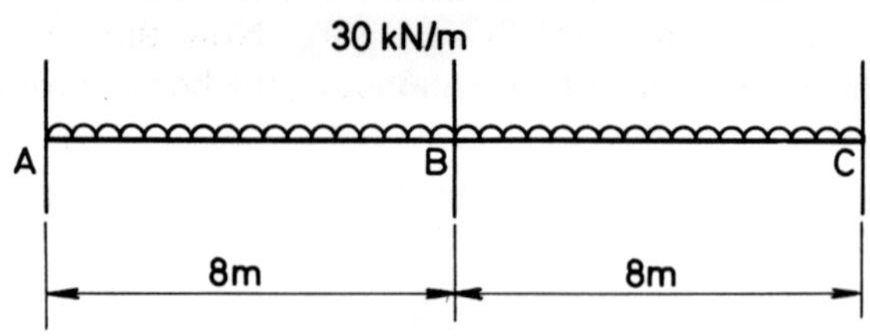

Figure 4.10. Problem 4.2

4.3. The two-storey frame with rigid joints, shown in Fig. 4.11, carries lateral loads as shown. Number the displacements as shown in the figure and assemble a stiffness matrix neglecting axial strains. Calculate the nodal displacements and the matrix of member forces.

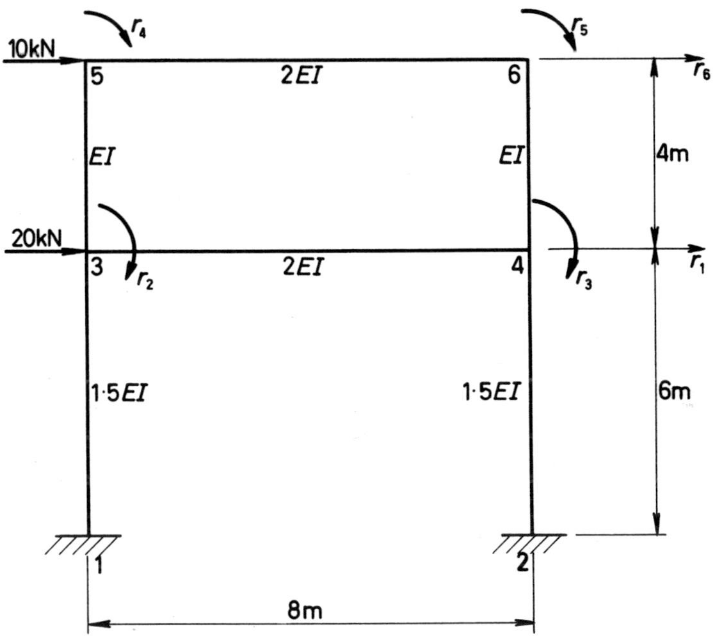

Figure 4.11. Problem 4.3

4.4. The frame shown in Fig. 4.12 has inclined columns and encastré supports. The members all have the same uniform *EI*. Calculate the bending moments in the column AB.

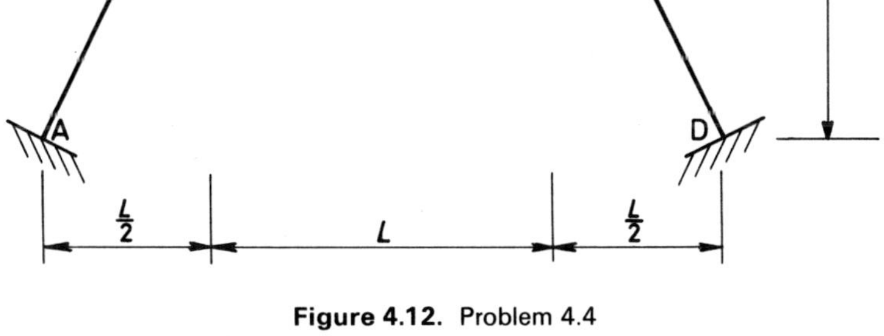

Figure 4.12. Problem 4.4

4.5. In the frame shown in Fig. 4.13 the beams GH and DE may be considered to have infinite flexural rigidity, whereas the columns and beam EF have $EI = 1000\ \text{kNm}^2$. Neglect axial strains and assemble a stiffness matrix for the structure.

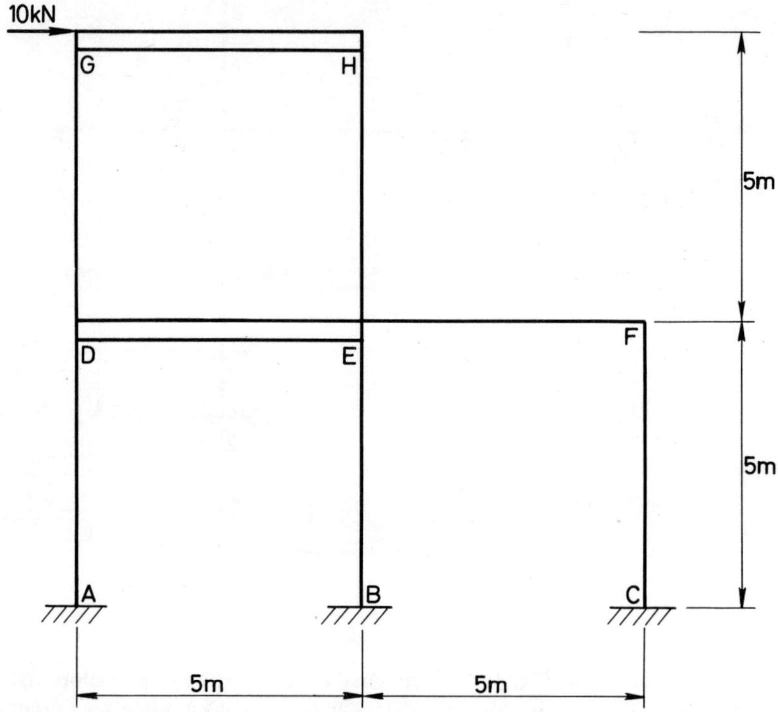

Figure 4.13. Problem 4.5

Hence, determine the horizontal deflection of the beam GH. If, for rapid analysis, the beam EF is also considered to have $EI = \infty$, what would be the error in the calculated deflection?

5. Additional Topics in the Stiffness Method

5.1. Non-prismatic Member Stiffnesses

The effects of varying cross-section of a beam element on the end flexibilities, and end rotations under load, were discussed in section 3.1, and a general method of computation based on the use of Simpson's rule was outlined. The use of the stiffness method with non-prismatic elements requires a knowledge of the direct and cross stiffnesses, and fixed end moments. These can be obtained once the flexibilities, and the end rotations due to the applied load, are known.

Consider a non-prismatic beam, Fig. 5.1a, carrying an applied load producing a bending moment diagram m_0 when ends 1 and 2 are simply-supported. The member flexibilities and end rotations due to applied load are

$$\left.\begin{aligned} f_{11} &= \int_0^l \frac{m_1^2 \, dx}{EI} \\ f_{22} &= \int_0^l \frac{m_2^2 \, dx}{EI} \\ f_{12} = f_{21} &= \int_0^l \frac{m_1 m_2 \, dx}{EI} \\ f_{01} &= \int_0^l \frac{m_0 m_1 \, dx}{EI} \\ f_{02} &= \int_0^l \frac{m_0 m_2 \, dx}{EI} \end{aligned}\right\} \tag{5.1}$$

Combining (e) and (f), Fig. 5.1, in the form (e)$-$(f) $\times f_{21}/f_{22}$, will remove the rotation at end 2. The resulting moment at end 1 is unity and the resulting rotation at end 1 is

$$f_{11} - f_{12}\frac{f_{21}}{f_{22}}$$

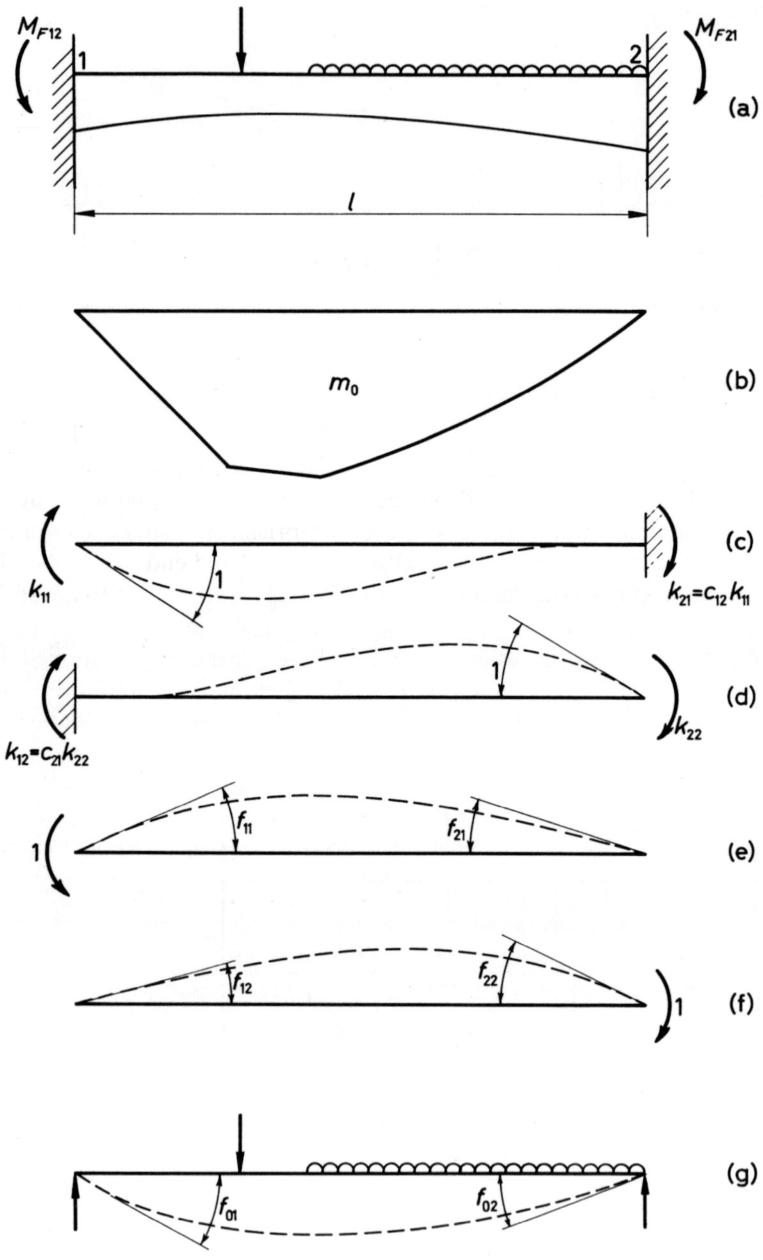

Figure 5.1. Non-prismatic beam; end flexibilities and stiffnesses

Thus $$k_{11} = 1\Big/\left(f_{11} - f_{12}\frac{f_{21}}{f_{22}}\right) = \frac{f_{22}}{f_{11}f_{22} - f_{12}^2} \tag{5.2}$$

The corresponding moment at end 2 is f_{12}/f_{22} and thus, the carry-over factor from 1 to 2 is

$$C_{12} = f_{12}/f_{22} \tag{5.3}$$

It follows that

$$k_{22} = \frac{f_{11}}{f_{11}f_{22} - f_{12}^2} \tag{5.4}$$

and $$C_{21} = f_{12}/f_{11} \tag{5.5}$$

The results expressed in eqns. (5.2) to (5.5) could, of course, have been obtained simply by inverting the flexibility matrix to obtain the stiffness matrix, i.e.,

$$\mathbf{F} = \begin{bmatrix} f_{11} & f_{12} \\ f_{21} & f_{22} \end{bmatrix} \tag{5.6}$$

and $$\mathbf{K} = \mathbf{F}^{-1} = \begin{bmatrix} \left(\dfrac{f_{22}}{f_{11}f_{22} - f_{12}^2}\right) & -\left(\dfrac{f_{12}}{f_{11}f_{22} - f_{12}^2}\right) \\ -\left(\dfrac{f_{21}}{f_{11}f_{22} - f_{12}^2}\right) & \left(\dfrac{f_{11}}{f_{11}f_{22} - f_{12}^2}\right) \end{bmatrix} \tag{5.7}$$

Stiffnesses for remote end pinned. These may be obtained directly from Figs. 5.1e and f,

$$\left.\begin{aligned} k_{11} &= 1/f_{11} \\ k_{22} &= 1/f_{22} \end{aligned}\right. \tag{5.8}$$

Fixed-end moments. These are the moments required to remove the end rotations f_{01} and f_{02}, Fig. 5.1g

$$\left.\begin{aligned} M_{F12} &= k_{11}f_{01} - c_{21}k_{22}f_{02} \\ M_{F21} &= k_{22}f_{02} - c_{12}k_{11}f_{01} \end{aligned}\right\} \tag{5.9}$$

Example 5.1. Calculate the direct and cross stiffnesses and fixed end moments for the beam AB of example 3.2, Fig. 3.3. The flexibilities and end rotations due to the applied load were computed in example 3.2. The stiffnesses and fixed-end moments may be obtained simply by substituting these values in eqns. (5.2) to (5.5) and (5.9), as follows:

$$k_{AA} = \frac{24{\cdot}11}{24{\cdot}11^2 - 16{\cdot}74^2} \times 10^6 = 80{\cdot}0825 \times 10^3 = k_{BB}$$

$$k_{AB} = k_{BA} = c_{BA}k_{BB} = c_{AB}k_{AA}$$

$$= \frac{16{\cdot}74}{24{\cdot}11^2 - 16{\cdot}74^2} \times 10^6 = 55{\cdot}6027 \times 10^3$$

$$M_{FAB} = 128{\cdot}6 \times 10^{-6}Wk_{AA} - 99{\cdot}08 \times 10^{-6}Wk_{AB}$$

$$= 4{\cdot}789W$$

$$M_{FBA} = 99{\cdot}08 \times 10^{-6}Wk_{BB} - 128{\cdot}6 \times 10^{-6}Wk_{BA}$$

$$= 0{\cdot}7841W$$

5.2. Influence of Shear

Effect of shear on member stiffnesses. The effect of shear forces on member flexibilities was discussed in section 3.2, and for non-prismatic members equations of type (3.15) are used for the evaluation of flexibilities. The procedure of section 5.1 may then be used to determine the member stiffnesses. For uniform cross-section beams the flexibilities may be calculated using eqns. (3.17) and (3.18), then, using eqns. (5.2) to (5.5),

$$k_{11} = \frac{(l/3EI) + (f/lGA)}{[(l/3EI) + (f/lGA)]^2 - [(l/6EI) - (f/lGA)]^2} \tag{5.10}$$

Putting

$$g = \frac{6fEI}{l^2GA} \tag{5.11}$$

then

$$k_{11} = \frac{2EI}{l}\left(\frac{2 + g}{1 + 2g}\right) = k_{22} \tag{5.12}$$

also,

$$c_{12} = c_{21} = \frac{(l/6EI) - (f/lGA)}{(l/3EI) + (f/lGA)}$$

$$= \left(\frac{1 - g}{2 + g}\right) \tag{5.13}$$

or

$$k_{12} = k_{21} = c_{21}k_{22} = c_{12}k_{11}$$

$$= \frac{2EI}{l}\left(\frac{1 - g}{1 + 2g}\right) \tag{5.14}$$

The form factor f required in eq. (5.11) may be obtained from Table 3.1 for the cross-sections listed, or from eq. (3.16) in other cases.

Effect of shear on fixed-end moments. Equations (5.9) may be regarded as general expressions for calculating fixed-end moments. The effect of shear on the stiffness coefficients in eq. (5.9) is obtained from eqns. (5.12) and (5.14). In considering the effect of shear on fixed-end moments due to transverse loading it is convenient to consider separately the effects of symmetrical and unsymmetrical loading.

(a) symmetrical loading:

$$f_{01} = f_{02} = f_0$$

therefore,

$$M_{F12} = \frac{2EI}{l}\left(\frac{2+g}{1+2g}\right)f_0\left[1 - \left(\frac{1-g}{2+g}\right)\right]$$

$$= \frac{2EI}{l}f_0 = M_{F21} \tag{5.15}$$

hence the fixed-end moments are independent of g for symmetrical loading.

Now

$$f_0 = f_{01} = \int_0^l \frac{m_0 m_1 \, \mathrm{d}x}{EI} + f\int_0^l \frac{v_0 v_1 \, \mathrm{d}x}{GA}$$

$$= \int_0^l \frac{m_0 m_1 \, \mathrm{d}x}{EI} + \frac{f v_1}{GA}\int_0^l v_0 \, \mathrm{d}x \tag{5.16}$$

since v_1 is constant (Fig. 3.5).

Now the equilibrium condition for the beam, in the transverse direction, leads to

$$\int_0^l v_0 \, \mathrm{d}x = 0 \tag{5.17}$$

Hence, the fixed-end moments are unaffected by shear for symmetrical loading.

(b) unsymmetrical loading:
The second integral in eq. (5.16) is always zero because of condition eq. (5.17), thus the end slopes are unaffected by shear and hence

$$\left.\begin{aligned} f_{01} &= \int_0^l \frac{m_0 m_1 \, \mathrm{d}x}{EI} \\ \text{and} \quad f_{02} &= \int_0^l \frac{m_0 m_2 \, \mathrm{d}x}{EI} \end{aligned}\right\} \tag{5.18}$$

Substitution of f_{01} and f_{02} from eq. (5.18) in eq. (5.9), using the modified stiffnesses from eqns. (5.10) to (5.14), yields the fixed-end moments.

Example 5.2. Derive expressions for the fixed-end moments for a uniform beam AB, Fig. 5.2, carrying a load W.

$$f_{01} = \frac{1}{EI}\int_0^a \frac{Wbx}{l}(1 - x/l)\,\mathrm{d}x + \frac{1}{EI}\int_0^b \frac{Wax}{l}\left(\frac{x}{l}\right)\mathrm{d}x$$

$$= \frac{Wab}{6l^2EI}(a^2 + 2b^2 + 3ab)$$

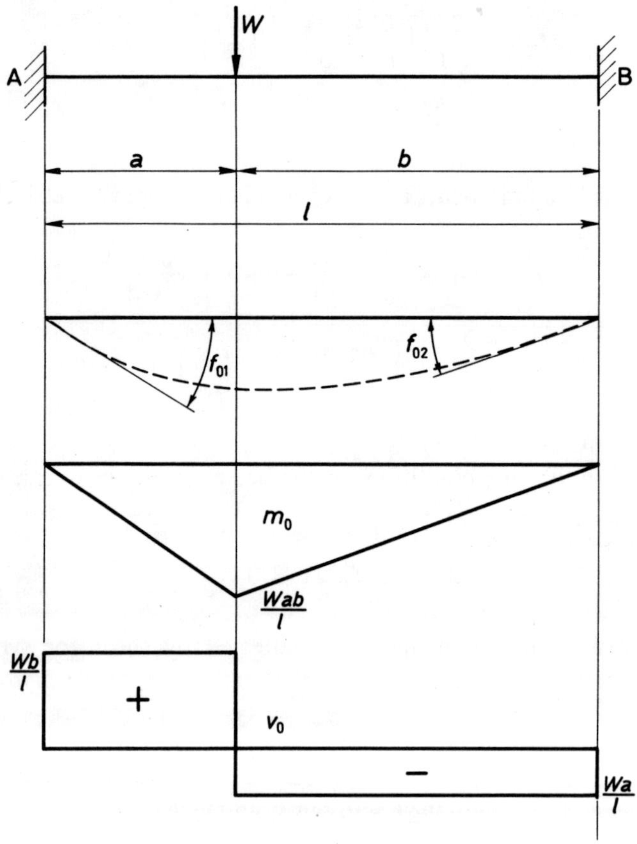

Figure 5.2. Example 5.2

and similarly,

$$f_{02} = \frac{Wab}{6l^2EI}(b^2 + 2a^2 + 3ab)$$

Substituting in eq. (5.9), and using eqns. (5.12) and (5.14),

$$M_{\mathrm{FAB}} = \frac{2EI}{l}\left(\frac{2+g}{1+2g}\right)$$

$$\times\left[\frac{Wab}{6l^2EI}(a^2+2b^2+3ab) - \left(\frac{1-g}{2+g}\right)\frac{Wab}{6l^2EI}(b^2+2a^2+3ab)\right]$$

$$= \frac{Wab}{l^2}\left(\frac{b+lg}{1+2g}\right)$$

and

$$M_{\mathrm{FBA}} = \frac{Wab}{l^2}\left(\frac{a+lg}{1+2g}\right)$$

5.3. Space Frame Analysis

The analysis of plane frame problems by the stiffness method has been accomplished so far using the simple, three-degree of freedom, stiffness matrix of eq. (4.2). This matrix contains all the information required for a stiffness analysis of a plane structure in which the applied loads are carried predominantly by flexure of the members. There are, of course, many other structures for which rather more complex stiffness matrices are required and at this stage it is appropriate to derive a general stiffness matrix of a straight, uniform, member suitable for the analysis of structures such as space frames with rigid joints and grillages.

The stiffness matrix will be developed first of all using the natural coordinate system of the member; then we shall transform the matrix into a more general structure coordinate system. This latter matrix will be more suitable for dealing with structures in which the members are inclined at various angles to an overall coordinate system.

Figure 5.3 shows a coordinate system for a typical member 1 2. The stiffness matrix may be developed, column by column, by giving a unit value to each of the twelve nodal degrees of freedom in turn. The elements in the stiffness matrix are then simply the forces acting at the nodes in the appropriate directions. The first three columns of the matrix are shown in Fig. 5.3. It is found that this particular ordering of the forces and displacements gives a compact stiffness matrix in diagonal submatrix form, as in eq. 5.19.

$$\begin{bmatrix} \mathbf{S}_Y \\ \mathbf{S}_Z \\ \mathbf{S}_T \\ \mathbf{S}_N \end{bmatrix} = \begin{bmatrix} \mathbf{k}_Y & & & \\ & \mathbf{k}_Z & & \\ & & \mathbf{k}_T & \\ & & & \mathbf{k}_N \end{bmatrix} \begin{bmatrix} \mathbf{r}_Y \\ \mathbf{r}_Z \\ \mathbf{r}_T \\ \mathbf{r}_N \end{bmatrix} \tag{5.19}$$

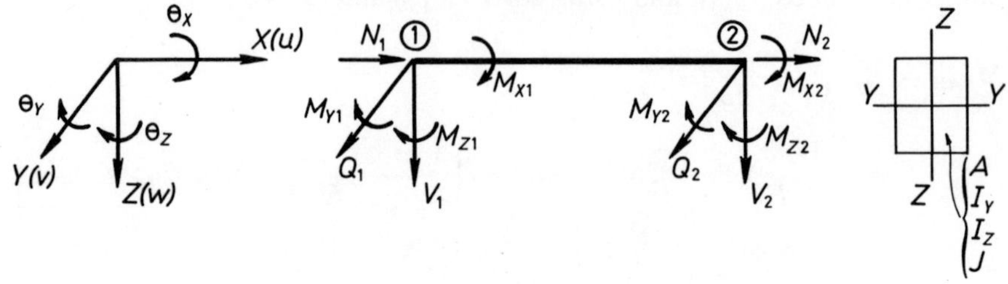

		$\theta_{Y1}=1$	$\theta_{Y2}=1$	$w_1=1$			
M_{Y1}		$4EI_y/l$	$2EI_y/l$	$6EI_y/l^2$			θ_{Y1}
M_{Y2}		$2EI_y/l$	$4EI_y/l$	$6EI_y/l^2$			θ_{Y2}
V_1		$6EI_y/l^2$	$6EI_y/l^2$	$12EI_y/l^3$			w_1
V_2		$-6EI_y/l^2$	$-6EI_y/l^2$	$-12EI_y/l^3$			w_2
M_{Z1}		0	0	0			θ_{Z1}
M_{Z2}	=	0	0	0			θ_{Z2}
Q_1		0	0	0			v_1
Q_2		0	0	0			v_2
M_{X1}		0	0	0			θ_{X1}
M_{X2}		0	0	0			θ_{X2}
N_1		0	0	0			u_1
N_2		0	0	0			u_2

Figure 5.3. Coordinate system and part stiffness matrix for space frame member

$$\left.\begin{aligned}
\mathbf{S}_Y &= \begin{bmatrix} M_{Y1} \\ M_{Y2} \\ V_1 \\ V_2 \end{bmatrix} & \mathbf{r}_Y &= \begin{bmatrix} \theta_{Y1} \\ \theta_{Y2} \\ w_1 \\ w_2 \end{bmatrix} \\
\mathbf{S}_Z &= \begin{bmatrix} M_{Z1} \\ M_{Z2} \\ Q_1 \\ Q_2 \end{bmatrix} & \mathbf{r}_Z &= \begin{bmatrix} \theta_{Z1} \\ \theta_{Z2} \\ v_1 \\ v_2 \end{bmatrix} \\
\mathbf{S}_T &= \begin{bmatrix} M_{X1} \\ M_{X2} \end{bmatrix} & \mathbf{r}_T &= \begin{bmatrix} \theta_{X1} \\ \theta_{X2} \end{bmatrix} \\
\mathbf{S}_N &= \begin{bmatrix} N_1 \\ N_2 \end{bmatrix} & \mathbf{r}_N &= \begin{bmatrix} u_1 \\ u_2 \end{bmatrix}
\end{aligned}\right\} \tag{5.20}$$

$$\mathbf{k}_Y = \frac{EI_Y}{l}\begin{bmatrix} 4 & 2 & 6/l & -6/l \\ 2 & 4 & 6/l & -6/l \\ 6/l & 6/l & 12/l^2 & -12/l^2 \\ -6/l & -6/l & -12/l^2 & 12/l^2 \end{bmatrix} \tag{5.21}$$

$$\mathbf{k}_Z = \frac{EI_Z}{l}\begin{bmatrix} 4 & 2 & 6/l & -6/l \\ 2 & 4 & 6/l & -6/l \\ 6/l & 6/l & 12/l^2 & -12/l^2 \\ -6/l & -6/l & -12/l^2 & 12/l^2 \end{bmatrix} \tag{5.22}$$

$$\mathbf{k}_T = \frac{GJ}{l}\begin{bmatrix} 1 & -1 \\ -1 & 1 \end{bmatrix} \tag{5.23}$$

$$\mathbf{k}_N = \frac{EA}{l}\begin{bmatrix} 1 & -1 \\ -1 & 1 \end{bmatrix} \tag{5.24}$$

The stiffness matrix given in eq. (5.19), its submatrices being defined in eqns. (5.21) to (5.24), may now be used as a stiffness matrix for a general member of a three-dimensional rigid frame. The transformation necessary to relate the stiffnesses in local, member, coordinates to those in global, structure, coordinates may be carried out using eq. (4.8),

$$\mathbf{K} = \mathbf{A}^T\mathbf{k}\mathbf{A}$$

in which the transformation is carried out for the entire structure in one operation. As we saw in chapter 4, this procedure makes heavy demands on computer storage and it is usually more economical to assemble the stiffness matrix directly by inspection or by the use of code numbers. This alternative process does, however, demand that the individual member stiffnesses are available in global coordinates and thus the transformations, where necessary, are carried out on the individual member stiffness matrices before commencing the assembly of the structure stiffness matrix.

We shall now examine the transformation of the member stiffnesses of eq. (5.19) for the general case of the member inclined to rectangular axes in space.

Since the compact form of matrix (5.19) will be lost on transformation, we may choose a more convenient order of the nodal forces and re-arrange the rows and columns of the stiffness matrix as in (5.25).

$$
\begin{bmatrix} M_{X1} \\ M_{Y1} \\ M_{Z1} \\ N_1 \\ Q_1 \\ V_1 \\ \hline M_{X2} \\ M_{Y2} \\ M_{Z2} \\ N_2 \\ Q_2 \\ V_2 \end{bmatrix}
=
\left[\begin{array}{cccccc|cccccc}
\frac{GJ}{l} & 0 & 0 & 0 & 0 & 0 & -\frac{GJ}{l} & 0 & 0 & 0 & 0 & 0 \\
0 & \frac{4EI_Y}{l} & 0 & 0 & 0 & \frac{6EI_Y}{l^2} & 0 & \frac{2EI_Y}{l} & 0 & 0 & 0 & -\frac{6EI_Y}{l^2} \\
0 & 0 & \frac{4EI_Z}{l} & 0 & \frac{6EI_Z}{l^2} & 0 & 0 & 0 & \frac{2EI_Z}{l} & 0 & -\frac{6EI_Z}{l^2} & 0 \\
0 & 0 & 0 & \frac{EA}{l} & 0 & 0 & 0 & 0 & 0 & -\frac{EA}{l} & 0 & 0 \\
0 & 0 & \frac{6EI_Z}{l^2} & 0 & \frac{12EI_Z}{l^3} & 0 & 0 & 0 & \frac{6EI_Z}{l^2} & 0 & -\frac{12EI_Z}{l^3} & 0 \\
0 & \frac{6EI_Y}{l^2} & 0 & 0 & 0 & \frac{12EI_Y}{l^3} & 0 & \frac{6EI_Y}{l^2} & 0 & 0 & 0 & -\frac{12EI_Y}{l^3} \\
\hline
-\frac{GJ}{l} & 0 & 0 & 0 & 0 & 0 & \frac{GJ}{l} & 0 & 0 & 0 & 0 & 0 \\
0 & \frac{2EI_Y}{l} & 0 & 0 & 0 & \frac{6EI_Y}{l^2} & 0 & \frac{4EI_Y}{l} & 0 & 0 & 0 & -\frac{6EI_Y}{l^2} \\
0 & 0 & \frac{2EI_Z}{l} & 0 & \frac{6EI_Z}{l^2} & 0 & 0 & 0 & \frac{4EI_Z}{l} & 0 & -\frac{6EI_Z}{l^2} & 0 \\
0 & 0 & 0 & -\frac{EA}{l} & 0 & 0 & 0 & 0 & 0 & \frac{EA}{l} & 0 & 0 \\
0 & 0 & -\frac{6EI_Z}{l^2} & 0 & -\frac{12EI_Z}{l^3} & 0 & 0 & 0 & -\frac{6EI_Z}{l^2} & 0 & \frac{12EI_Z}{l^3} & 0 \\
0 & -\frac{6EI_Y}{l^2} & 0 & 0 & 0 & -\frac{12EI_Y}{l^3} & 0 & -\frac{6EI_Y}{l^2} & 0 & 0 & 0 & \frac{12EI_Y}{l^3}
\end{array}\right]
\begin{bmatrix} \theta_{X1} \\ \theta_{Y1} \\ \theta_{Z1} \\ u_1 \\ v_1 \\ w_1 \\ \hline \theta_{X2} \\ \theta_{Y2} \\ \theta_{Z2} \\ u_2 \\ v_2 \\ w_2 \end{bmatrix}
\tag{5.25}
$$

Equation (5.25) may be written in partitioned form as

$$\begin{bmatrix} \mathbf{S}_1 \\ \mathbf{S}_2 \end{bmatrix} = \begin{bmatrix} \mathbf{k}_{11} & \mathbf{k}_{12} \\ \mathbf{k}_{21} & \mathbf{k}_{22} \end{bmatrix} \begin{bmatrix} \mathbf{r}_1 \\ \mathbf{r}_2 \end{bmatrix} \tag{5.26}$$

A new set of coordinate axes $\bar{X}$, $\bar{Y}$, $\bar{Z}$ is defined by rotation of the original X, Y, Z set. The relationships between the nodal forces in the new and original coordinate systems is given for either end by

$$\begin{bmatrix} \bar{M}_{\bar{X}} \\ \bar{M}_{\bar{Y}} \\ \bar{M}_{\bar{Z}} \\ \bar{N} \\ \bar{Q} \\ \bar{V} \end{bmatrix} = \begin{bmatrix} \lambda_{\bar{X}X} & \lambda_{\bar{X}Y} & \lambda_{\bar{X}Z} & 0 & 0 & 0 \\ \lambda_{\bar{Y}X} & \lambda_{\bar{Y}Y} & \lambda_{\bar{Y}Z} & 0 & 0 & 0 \\ \lambda_{\bar{Z}X} & \lambda_{\bar{Z}Y} & \lambda_{\bar{Z}Z} & 0 & 0 & 0 \\ 0 & 0 & 0 & \lambda_{\bar{X}X} & \lambda_{\bar{X}Y} & \lambda_{\bar{X}Z} \\ 0 & 0 & 0 & \lambda_{\bar{Y}X} & \lambda_{\bar{Y}Y} & \lambda_{\bar{Y}Z} \\ 0 & 0 & 0 & \lambda_{\bar{Z}X} & \lambda_{\bar{Z}Y} & \lambda_{\bar{Z}Z} \end{bmatrix} \begin{bmatrix} M_X \\ M_Y \\ M_Z \\ N \\ Q \\ V \end{bmatrix}$$

where $\lambda_{\bar{X}X} = \cos \bar{X}OX$, etc.

or

$$\bar{\mathbf{S}} = \boldsymbol{\lambda}\mathbf{S} \tag{5.27}$$

Using the principle of contragredience, eq. (1.39), the displacement transformation is

$$\mathbf{r} = \boldsymbol{\lambda}^T\bar{\mathbf{r}} \tag{5.28}$$

or

$$\bar{\mathbf{r}} = (\boldsymbol{\lambda}^T)^{-1}\mathbf{r} \tag{5.29}$$

Now it can be shown that $(\boldsymbol{\lambda}^T)^{-1} = \boldsymbol{\lambda}$, this being an 'orthogonal' transformation, hence

$$\bar{\mathbf{r}} = \boldsymbol{\lambda}\mathbf{r} \tag{5.30}$$

Applying eqns. (4.5) and (4.8)

$$\bar{\mathbf{k}} = \boldsymbol{\lambda}\mathbf{k}\boldsymbol{\lambda}^T \tag{5.31}$$

The complete stiffness matrix in the $\bar{X}$, $\bar{Y}$, $\bar{Z}$ system is conveniently expressed in partitioned form

$$\begin{bmatrix} \bar{\mathbf{S}}_1 \\ \bar{\mathbf{S}}_2 \end{bmatrix} = \begin{bmatrix} \boldsymbol{\lambda} & \mathbf{0} \\ \mathbf{0} & \boldsymbol{\lambda} \end{bmatrix} \begin{bmatrix} \mathbf{k}_{11} & \mathbf{k}_{12} \\ \mathbf{k}_{21} & \mathbf{k}_{22} \end{bmatrix} \begin{bmatrix} \boldsymbol{\lambda}^T & \mathbf{0} \\ \mathbf{0} & \boldsymbol{\lambda}^T \end{bmatrix} \begin{bmatrix} \bar{\mathbf{r}}_1 \\ \bar{\mathbf{r}}_2 \end{bmatrix}$$

$$= \begin{bmatrix} \boldsymbol{\lambda}\mathbf{k}_{11}\boldsymbol{\lambda}^T & \boldsymbol{\lambda}\mathbf{k}_{12}\boldsymbol{\lambda}^T \\ \boldsymbol{\lambda}\mathbf{k}_{21}\boldsymbol{\lambda}^T & \boldsymbol{\lambda}\mathbf{k}_{22}\boldsymbol{\lambda}^T \end{bmatrix} \begin{bmatrix} \bar{\mathbf{r}}_1 \\ \bar{\mathbf{r}}_2 \end{bmatrix} \tag{5.32}$$

which is the transformed version of expression (5.26).

The stiffness matrix for an inclined member in a plane frame may be obtained from eq. (5.32) by suppressing appropriate rows and columns. Assuming that the member lies in the XZ plane, Fig. 5.3, inclined at angle θ to the $\bar{X}$ axis, and is free from torsion, the conditions to be introduced are

$$\theta_X = \theta_Z = v = 0 \tag{5.33}$$

and

$$\begin{aligned} &\lambda_{\bar{X}X} = \cos\theta, \quad \lambda_{\bar{Y}Y} = 1, \quad \lambda_{\bar{Z}Z} = \cos\theta \\ &\lambda_{\bar{X}Y} = 0, \quad \lambda_{\bar{Y}X} = 0, \quad \lambda_{\bar{Z}X} = -\sin\theta \\ &\lambda_{\bar{X}Z} = \sin\theta, \quad \lambda_{\bar{Y}Z} = 0, \quad \lambda_{\bar{Z}Y} = 0 \end{aligned} \tag{5.34}$$

Hence,

$$\boldsymbol{\lambda} = \begin{bmatrix} 1 & 0 & 0 \\ 0 & \cos\theta & \sin\theta \\ 0 & -\sin\theta & \cos\theta \end{bmatrix} \tag{5.35}$$

and from eq. (5.32), with $EI_Y = EI$, we obtain the transformed stiffness matrix of eq. (5.36).

Example 5.3. The structure shown in Fig. 5.4 is composed of three rigid frames with inclined rafters, orientated in plan at angles of 60°, and rigidly connected at the apex. The problem is taken to be that of constructing a stiffness matrix for the structure. The local, XYZ, and global, $\bar{X}\bar{Y}\bar{Z}$, co-ordinate systems are as shown in the figure. In order to save space, only an outline of the matrix assembly will be given, using the code number method.

Neglecting any possible symmetry in applied loading, the structure has $7 \times 6 = 42$ degrees of freedom. The treatment of symmetry, outlined in section 4.4, would be appropriate to this problem but this will be left as an exercise for the reader.

All the members have the same uniform section with properties EA, EI_Y, EI_Z, and GJ. The individual member stiffness matrices $\mathbf{k}_{12}$, $\mathbf{k}_{34}$, etc., in local coordinates may be formed using eq. (5.25), and this is a straightforward exercise. Before the code number method can be applied to generate the structure stiffness matrix, the individual member stiffnesses must be transformed into the global coordinate system. This is achieved using eq. (5.32) and it is convenient to first compile a table of direction cosines from which the matrix $\boldsymbol{\lambda}$ appropriate to a particular member may be found. As an illustration, the direction cosines for three of the members are tabulated below:

Member	$\lambda_{\bar{X}X}$	$\lambda_{\bar{X}Y}$	$\lambda_{\bar{X}Z}$	$\lambda_{\bar{Y}X}$	$\lambda_{\bar{Y}Y}$	$\lambda_{\bar{Y}Z}$	$\lambda_{\bar{Z}X}$	$\lambda_{\bar{Z}Y}$	$\lambda_{\bar{Z}Z}$
1–2	0	$-\sqrt{3}/2$	$\frac{1}{2}$	0	$\frac{1}{2}$	$\sqrt{3}/2$	-1	0	0
3–4	0	$-\sqrt{3}/2$	$-\frac{1}{2}$	0	$-\frac{1}{2}$	$\sqrt{3}/2$	-1	0	0
5–6	0	0	1	0	1	0	-1	0	0

Direction cosines for example 5.3

$$\bar{\mathbf{k}} = \begin{bmatrix} \frac{4EI}{l} & & & & & \text{Symmetric} \\ \frac{6EI}{l^2}\sin\theta & \frac{12EI}{l^3}\sin^2\theta + \frac{EA}{l}\cos^2\theta & & & & \\ \frac{6EI}{l^2}\cos\theta & \frac{12EI}{l^3}\cos\theta\sin\theta - \frac{EA}{l}\cos\theta\sin\theta & \frac{12EI}{l^3}\cos^2\theta + \frac{EA}{l}\sin^2\theta & & & \\ \frac{2EI}{l} & \frac{6EI}{l^2}\sin\theta & \frac{6EI}{l^2}\cos\theta & \frac{4EI}{l} & & \\ -\frac{6EI}{l^2}\sin\theta & -\frac{12EI}{l^3}\sin^2\theta - \frac{EA}{l}\cos^2\theta & -\frac{12EI}{l^3}\sin\theta\cos\theta + \frac{EA}{l}\sin\theta\cos\theta & -\frac{6EI}{l^2}\sin\theta & \frac{12EI}{l^3}\sin^2\theta + \frac{EA}{l}\cos^2\theta & \\ -\frac{6EI}{l^2}\cos\theta & -\frac{12EI}{l^3}\sin\theta\cos\theta + \frac{EA}{l}\sin\theta\cos\theta & -\frac{12EI}{l^3}\cos^2\theta - \frac{EA}{l}\sin^2\theta & -\frac{6EI}{l^2}\cos\theta & \frac{12EI}{l^3}\sin\theta\cos\theta - \frac{EA}{l}\sin\theta\cos\theta & \frac{12EI}{l^3}\cos^2\theta + \frac{EA}{l}\sin^2\theta \end{bmatrix} \tag{5.36}$$

The transformation of the member stiffnesses is then carried out in accordance with eq. (5.32), for example,

$$\bar{\mathbf{k}}_{1-2} = \begin{bmatrix} \boldsymbol{\lambda}_{1-2} & \mathbf{0} \\ \mathbf{0} & \boldsymbol{\lambda}_{1-2} \end{bmatrix} [\mathbf{k}_{1-2}] \begin{bmatrix} \boldsymbol{\lambda}_{1-2}{}^T & \mathbf{0} \\ \mathbf{0} & \boldsymbol{\lambda}_{1-2}{}^T \end{bmatrix}$$

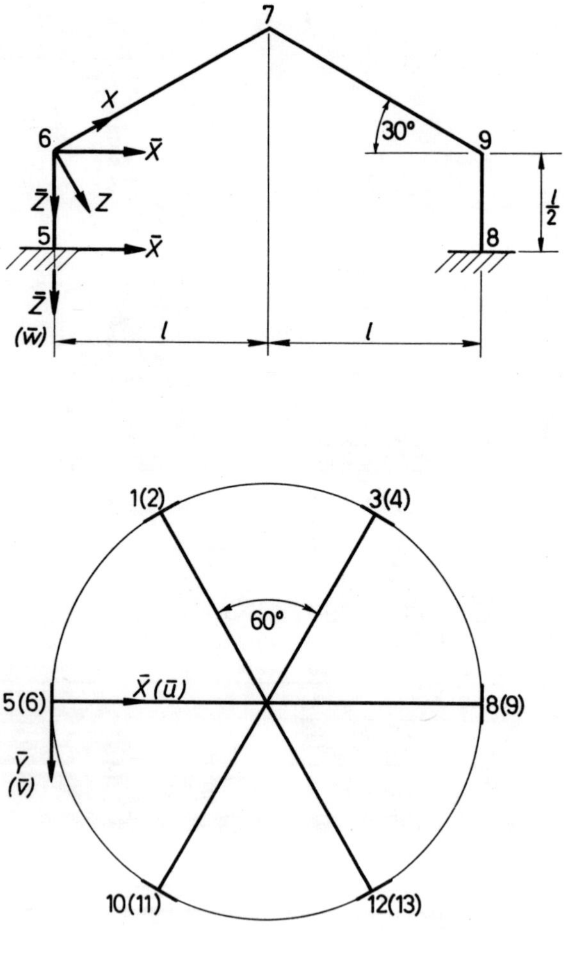

Figure 5.4. Example 5.3

Where $\mathbf{k}_{1-2}$, the stiffness matrix in local coordinates, is given in eq. (5.25), and

$$\boldsymbol{\lambda}_{1-2} = \left[\begin{array}{ccc|ccc} 0 & -\sqrt{3}/2 & \frac{1}{2} & 0 & 0 & 0 \\ 0 & \frac{1}{2} & \sqrt{3}/2 & 0 & 0 & 0 \\ -1 & 0 & 0 & 0 & 0 & 0 \\ \hline 0 & 0 & 0 & 0 & -\sqrt{3}/2 & \frac{1}{2} \\ 0 & 0 & 0 & 0 & \frac{1}{2} & \sqrt{3}/2 \\ 0 & 0 & 0 & -1 & 0 & 0 \end{array}\right]$$

If the nodal degrees of freedom are numbered, in the order $\theta_{\bar{X}}, \theta_{\bar{Y}}, \theta_{\bar{Z}}, \bar{u}, \bar{v}, \bar{w}$, at nodes 2, 4, 6, 7, 9, 11, and 13 in turn, then the code number for member 1–2 will be

0	0	0	0	0	0	1	2	3	4	5	6
$\theta_{\bar{X}}$	$\theta_{\bar{Y}}$	$\theta_{\bar{Z}}$	$\bar{u}$	$\bar{v}$	$\bar{w}$	$\theta_{\bar{X}}$	$\theta_{\bar{Y}}$	$\theta_{\bar{Z}}$	$\bar{u}$	$\bar{v}$	$\bar{w}$

Similarly, the code number for member 2–7 will be

1	2	3	4	5	6	19	20	21	22	23	24

The code numbers for the remaining members are written down by inspection in a similar way. The assembly of the structure stiffness matrix is then carried out by the process described in section 4.8.

5.4. Grillage Analysis

The structural grillage is a special case of the general space frame obtained by eliminating the 'in-plane' loading and 'in-plane' forces. For a horizontal grillage composed of straight, uniform, members subjected to transverse (vertical) loading only, the member stiffness matrix may be obtained by inspection from the general space frame member matrix (5.25), by the suppression of the rows and columns corresponding to M_Z, N, and Q.

Thus we obtain, for the typical grillage member shown in Fig. 5.5, with

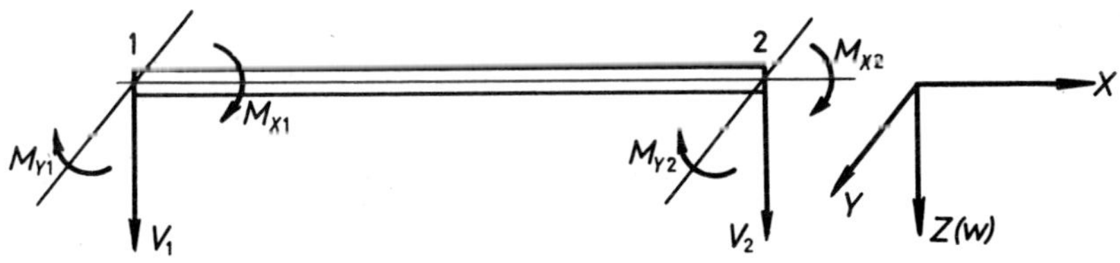

Figure 5.5. Coordinate system for grillage member

the X and Y axes lying in the plane of the grillage, a stiffness matrix, as in (5.37).

$$\begin{bmatrix} M_{X1} \\ M_{Y1} \\ V_1 \\ M_{X2} \\ M_{Y2} \\ V_2 \end{bmatrix} = \begin{bmatrix} \frac{GJ}{l} & 0 & 0 & -\frac{GJ}{l} & 0 & 0 \\ 0 & \frac{4EI}{l} & \frac{6EI}{l^2} & 0 & \frac{2EI}{l} & -\frac{6EI}{l^2} \\ 0 & \frac{6EI}{l^2} & \frac{12EI}{l^3} & 0 & \frac{6EI}{l^2} & -\frac{12EI}{l^3} \\ -\frac{GJ}{l} & 0 & 0 & \frac{GJ}{l} & 0 & 0 \\ 0 & \frac{2EI}{l} & \frac{6EI}{l^2} & 0 & \frac{4EI}{l} & -\frac{6EI}{l^2} \\ 0 & -\frac{6EI}{l^2} & -\frac{12EI}{l^3} & 0 & -\frac{6EI}{l^2} & \frac{12EI}{l^3} \end{bmatrix} \begin{bmatrix} \theta_{X1} \\ \theta_{Y1} \\ w_1 \\ \theta_{X2} \\ \theta_{Y2} \\ w_2 \end{bmatrix} \tag{5.37}$$

where $EI = EI_Y$.

The stiffness matrix, given in eq. (5.37), will suffice for all grillages in which the longitudinal and transverse members intersect at right-angles. For skew grillages, as in Fig. 5.6a, it will be necessary to adopt a transformation of the member stiffnesses to meet the general case. It will be convenient to direct the X or Y axis parallel to the longitudinal axis of the grillage; the stiffness matrix of eq. (5.37) will then apply directly to the longitudinal members, Fig. 5.6b. The stiffness matrix appropriate to the transverse members can then be expressed in the global system by a transformation. The axes are directed as shown in Fig. 5.6c.

The transformation matrix can be obtained from eq. (5.27) by inspection. Noting that the Z and $\bar{Z}$ axes are always perpendicular to the plane of the grillage, then the transformation matrix is given by

$$\boldsymbol{\lambda} = \begin{bmatrix} \lambda_{\bar{X}X} & \lambda_{\bar{X}Y} & 0 \\ \lambda_{\bar{Y}X} & \lambda_{\bar{Y}Y} & 0 \\ 0 & 0 & \lambda_{\bar{Z}Z} \end{bmatrix} = \begin{bmatrix} \cos\theta & \sin\theta & 0 \\ -\sin\theta & \cos\theta & 0 \\ 0 & 0 & 1 \end{bmatrix} \tag{5.38}$$

The transformed version of eq. (5.37) then becomes, in partitioned form,

$$\begin{bmatrix} \bar{\mathbf{S}}_1 \\ \bar{\mathbf{S}}_2 \end{bmatrix} = \begin{bmatrix} \boldsymbol{\lambda}\mathbf{k}_{11}\boldsymbol{\lambda}^T & \boldsymbol{\lambda}\mathbf{k}_{12}\boldsymbol{\lambda}^T \\ \boldsymbol{\lambda}\mathbf{k}_{21}\boldsymbol{\lambda}^T & \boldsymbol{\lambda}\mathbf{k}_{22}\boldsymbol{\lambda}^T \end{bmatrix} \begin{bmatrix} \bar{\mathbf{r}}_1 \\ \bar{\mathbf{r}}_2 \end{bmatrix}$$

which is given in expanded form in eq. (5.39).

$$
\begin{Bmatrix} \overline{M}_{\overline{X}1} \\ \overline{M}_{\overline{Y}1} \\ \bar{V}_1 \\ \overline{M}_{\overline{X}2} \\ \overline{M}_{\overline{Y}2} \\ \bar{V}_2 \end{Bmatrix}
=
\begin{bmatrix}
\frac{4EI}{l}\sin^2\theta - \frac{GJ}{l}\cos^2\theta & & & & & \\
\frac{4EI}{l}\sin\theta\cos\theta - \frac{GJ}{l}\sin\theta\cos\theta & \frac{4EI}{l}\cos^2\theta + \frac{GJ}{l}\sin^2\theta & & & \text{Symmetric} & \\
\frac{6EI}{l^2}\sin\theta & \frac{6EI}{l^2}\cos\theta & \frac{12EI}{l^3} & & & \\
\frac{2EI}{l}\sin^2\theta - \frac{GJ}{l}\cos^2\theta & \frac{2EI}{l}\sin\theta\cos\theta + \frac{GJ}{l}\sin\theta\cos\theta & \frac{6EI}{l^2}\sin\theta & \frac{4EI}{l}\sin^2\theta + \frac{GJ}{l}\cos^2\theta & & \\
\frac{2EI}{l}\sin\theta\cos\theta + \frac{GJ}{l}\sin\theta\cos\theta & \frac{2EI}{l}\cos^2\theta - \frac{GJ}{l}\sin^2\theta & \frac{6EI}{l^2}\cos\theta & \frac{4EI}{l}\sin\theta\cos\theta - \frac{GJ}{l}\sin\theta\cos\theta & \frac{4EI}{l}\cos^2\theta + \frac{GJ}{l}\sin^2\theta & \\
-\frac{6EI}{l^2}\sin\theta & -\frac{6EI}{l^2}\cos\theta & -\frac{12EI}{l^3} & -\frac{6EI}{l^2}\sin\theta & -\frac{6EI}{l^2}\cos\theta & \frac{12EI}{l^3}
\end{bmatrix}
\begin{Bmatrix} \bar{\theta}_{\overline{X}1} \\ \bar{\theta}_{\overline{Y}1} \\ \bar{w}_1 \\ \bar{\theta}_{\overline{X}2} \\ \bar{\theta}_{\overline{Y}2} \\ \bar{w}_2 \end{Bmatrix}
\qquad (5.39)
$$

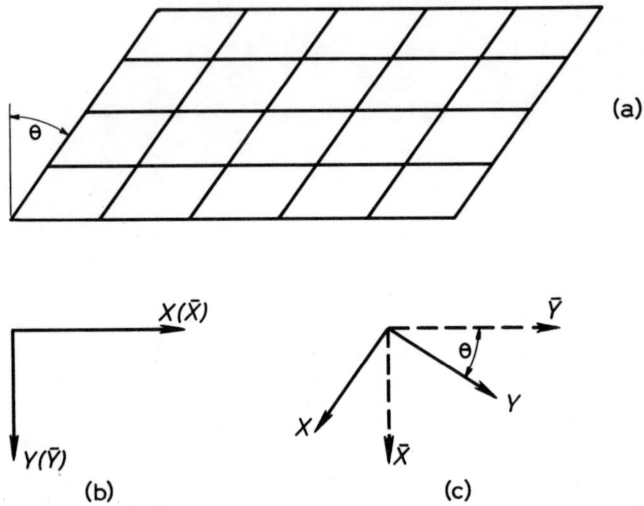

Figure 5.6. Skew grillage

Once the structure stiffness matrix has been assembled in global coordinates, and the equilibrium equations solved for a particular set of applied loads, the member forces are computed from equations of type (5.26).

If a transformation of stiffnesses is involved with a particular member then, from eqns. (5.26) and (5.28),

$$\mathbf{S} = \mathbf{k}\boldsymbol{\lambda}^T\bar{\mathbf{r}} \tag{5.40}$$

which has a similar form to eq. (4.17).

5.5. Banded Stiffness Matrices

A 'band' matrix is one in which all the non-zero terms lie on or close to the principal diagonal. The matrices of coefficients in the continuous beam analyses, sections 3.7 and 4.6, were found to be banded in form, eqns. (3.29) and (4.32).

Band matrices occur frequently in the stiffness method and the advances in computer analysis of structures in the last few years have been made possible largely by taking advantage of the band matrix. As we have already seen, in both flexibility and stiffness methods, the coefficient matrix is symmetric. Thus, the demands on computer storage can be very substantially reduced if, instead of storing the whole matrix, we store only a 'one-half' band.

Consider a two-bay frame with rigid joints, 30 storeys high, Fig. 5.7. Considering three degrees of freedom at each joint there are 270 unknowns in the equilibrium equations and if the entire, square, stiffness matrix is stored in the computer it will require $270^2 = 72{,}900$ registers. The half-band width of the stiffness matrix is 12 and thus the storage required for the half-band is only $12 \times 270 = 3240$ registers.

Methods of solution of band matrices will be considered in chapter 7, section 7.4. It is clearly desirable to keep the band width to a minimum and this is affected by the order in which the nodal displacements are numbered.

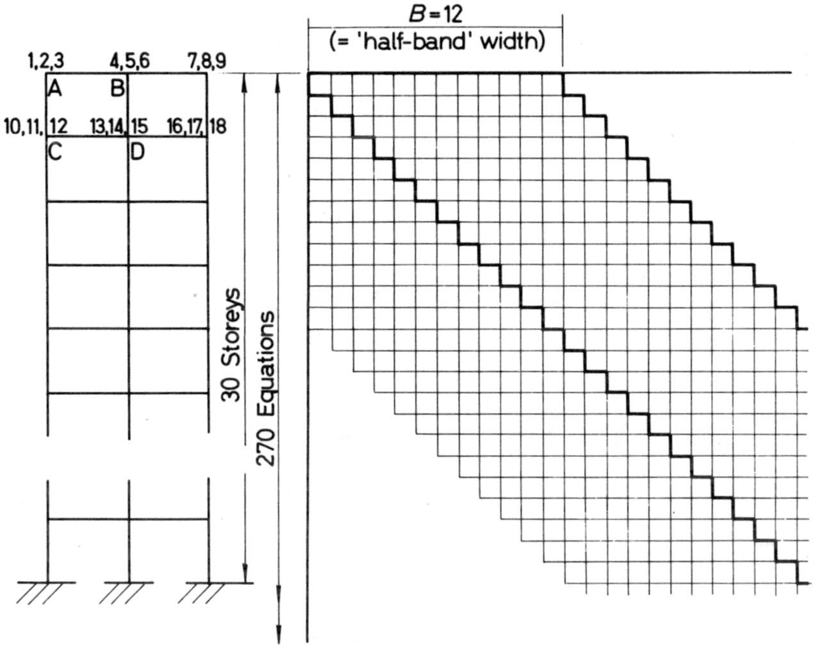

Figure 5.7. Banded stiffness matrix

Considering the rigid jointed frame in Fig. 5.7 and applying constraints to all the nodal degrees of freedom except number 1 at joint A, it is clear that an imposed unit displacement in the direction of 1 will require constraining forces at the nodes directly connected to node A, i.e., B and C. These constraining forces are, of course, the cross-stiffnesses appearing in the stiffness matrix and these forces are confined to the joints B and C. Thus, the non-zero terms in the first row of the stiffness matrix will be confined to the first 12 positions. This defines the half-band width as the maximum difference between the numbered degrees of freedom at the ends of any column + 1.

In fact, the numbering system shown in Fig. 5.7 produces the narrowest band possible for this structure. It is left as an exercise for the reader to show that, if the displacements are numbered from top to bottom down the joints of each column in turn, the half-band width will be 93.

5.6. Condensation of Stiffness Matrix

The complete structure stiffness matrix may refer to more degrees of freedom than actually required for the problem in hand. In this case, the stiffness matrix may be 'condensed' by eliminating the unwanted degrees of freedom. The resulting stiffness matrix expresses the stiffness of the structure in the limited number of displacements chosen. The process is essentially one of partial elimination of the unknown displacements.

Suppose we have a structure in which the displacements $\mathbf{r}_1$ constitute a set in which we are not interested, and $\mathbf{r}_2$ constitute a set in which we wish to establish a condensed stiffness matrix. Then we arrange to assemble the stiffness matrix in partitioned form, as in (5.41),

$$\begin{bmatrix} \mathbf{K}_1 & \mathbf{K}_3^T \\ \mathbf{K}_3 & \mathbf{K}_2 \end{bmatrix} \begin{bmatrix} \mathbf{r}_1 \\ \mathbf{r}_2 \end{bmatrix} = \begin{bmatrix} \mathbf{R}_1 \\ \mathbf{R}_2 \end{bmatrix} \tag{5.41}$$

that is, $$\mathbf{K}_1\mathbf{r}_1 + \mathbf{K}_3^T\mathbf{r}_2 = \mathbf{R}_1$$

and $$\mathbf{K}_3\mathbf{r}_1 + \mathbf{K}_2\mathbf{r}_2 = \mathbf{R}_2$$

eliminating $\mathbf{r}_1$,

$$(\mathbf{K}_2 - \mathbf{K}_3\mathbf{K}_1^{-1}\mathbf{K}_3^T)\mathbf{r}_2 = \mathbf{R}_2 - \mathbf{K}_3\mathbf{K}_1^{-1}\mathbf{R}_1$$

or $$\overline{\mathbf{K}}\mathbf{r}_2 = \overline{\mathbf{R}} \tag{5.42}$$

where $$\overline{\mathbf{K}} = (\mathbf{K}_2 - \mathbf{K}_3\mathbf{K}_1^{-1}\mathbf{K}_3^T) \tag{5.43}$$

and $$\overline{\mathbf{R}} = (\mathbf{R}_2 - \mathbf{K}_3\mathbf{K}_1^{-1}\mathbf{R}_1) \tag{5.44}$$

$\overline{\mathbf{K}}$ is the condensed stiffness matrix expressing the stiffness of the structure in the degrees of freedom $\mathbf{r}_2$, and $\overline{\mathbf{R}}$ is a modified load matrix.

Example 5.4. The stepped beam shown in Fig. 5.8 is $3l$ in length and has

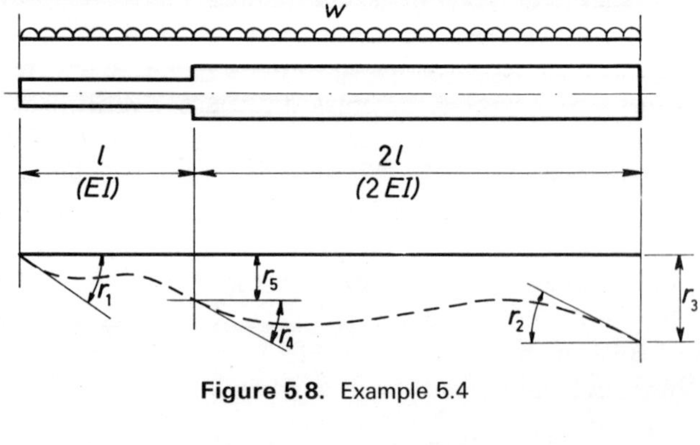

Figure 5.8. Example 5.4

flexural rigidities EI over length l and $2EI$ over length $2l$. A matrix condensation process will be used to determine the stiffness matrix of the beam in terms of the end displacements $r_{1,2,3}$.

The complete stiffness matrix, in the partitioned form of eq. (5.41), can be written down by inspection:

$$EI\left[\begin{array}{cc|ccc} \frac{4}{l}+\frac{8}{2l} & \frac{12}{4l^2}-\frac{6}{l^2} & \frac{2}{l} & \frac{4}{2l} & -\frac{12}{4l^2} \\ -\frac{3}{l^2} & \frac{12}{l^3}+\frac{24}{8l^3} & -\frac{6}{l^2} & \frac{12}{4l^2} & -\frac{24}{8l^3} \\ \hline \frac{2}{l} & -\frac{6}{l^2} & \frac{4}{l} & 0 & 0 \\ \frac{2}{l} & \frac{3}{l^2} & 0 & \frac{8}{2l} & -\frac{12}{4l^2} \\ -\frac{3}{l^2} & -\frac{3}{l^3} & 0 & -\frac{3}{l^2} & \frac{24}{8l^3} \end{array}\right]\left[\begin{array}{c} \mathbf{r}_4 \\ \mathbf{r}_5 \\ \hline \mathbf{r}_1 \\ \mathbf{r}_2 \\ \mathbf{r}_3 \end{array}\right] = wl\left[\begin{array}{c} \frac{l}{3}-\frac{l}{12} \\ \frac{1}{2}+1 \\ \hline \frac{l}{12} \\ -\frac{l}{3} \\ 1 \end{array}\right] \qquad (5.45)$$

From eq. (5.43)

$$\overline{\mathbf{K}} = \frac{EI}{l^3}\begin{bmatrix} 4l^2 & 0 & 0 \\ 0 & 4l^2 & -3l \\ 0 & -3l & 3 \end{bmatrix}$$

$$-\frac{EI}{l^3}\begin{bmatrix} 2l^2 & -6l \\ 2l^2 & 3l \\ -3l & -3 \end{bmatrix}\begin{bmatrix} \frac{5}{37l^2} & \frac{1}{37l} \\ \frac{1}{37l} & \frac{8}{111} \end{bmatrix}\begin{bmatrix} 2l^2 & 2l^2 & -3l \\ -6l & 3l & -3 \end{bmatrix}$$

$$= \frac{EI}{37l^3}\begin{bmatrix} 56l^2 & 34l^2 & -30l \\ 34l^2 & 92l^2 & -42l \\ -30l & -42l & 24 \end{bmatrix} \qquad (5.46)$$

From eq. (5.44)

$$\overline{\mathbf{R}} = wl\begin{bmatrix} \frac{l}{12} \\ -\frac{l}{3} \\ 1 \end{bmatrix} - \frac{1}{37}\begin{bmatrix} 2l^2 & -6l \\ 2l^2 & 3l \\ -3l & -3 \end{bmatrix}\begin{bmatrix} \frac{5}{l^2} & \frac{1}{l} \\ \frac{1}{l} & \frac{8}{3} \end{bmatrix}\begin{bmatrix} \frac{l}{4} \\ \frac{3}{2} \end{bmatrix} wl$$

$$\overline{\mathbf{R}} = \frac{wl}{444}\begin{bmatrix} 277l \\ -367l \\ 696 \end{bmatrix} \tag{5.47}$$

The matrix $\overline{\mathbf{R}}$ represents the 'fixed-end' effects of the applied loading on the non-prismatic beam.

Example 5.5. An eight-storey, single-bay, frame is shown in Fig. 5.9, and alongside is shown the stiffness matrix in which the non-zero elements are

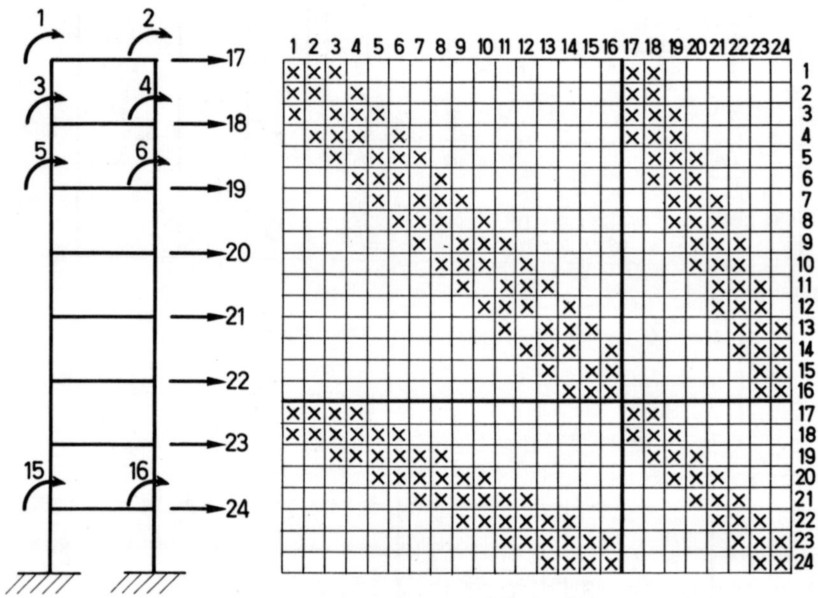

Figure 5.9. Example 5.5

shown by crosses. Neglecting axial displacements and numbering the nodal displacements as shown, the pattern of the stiffness matrix can be developed by inspection. The displacements have been numbered so that the lateral displacements of the storeys are concentrated into the bottom right-hand corner of the stiffness matrix.

Equation (5.43) may now be applied to produce a stiffness matrix which expresses the stiffness of the structure entirely in terms of the lateral displacements.

In structures with many degrees of freedom the inversion of matrix $\mathbf{K}_1$, eq. (5.43), may be excessively demanding in computer storage and advantage should be taken of the banded form of this matrix.[22]

5.7. Method of Substructures

We have seen that the banded form of the stiffness matrix should be exploited to obtain efficient solution processes and reduce demands on computer storage. If the band of non-zero coefficients in the stiffness matrix is uniform in width over the full matrix then this additional feature simplifies the programming of the solution. If the band width changes due to a change in layout at some point in the structure, Fig. 5.10a, then it may be convenient to consider the total structure divided into two, or more, substructures with common nodes. An alternative procedure, which may be attractive when the structure is not very large and computer space is available, is to introduce dummy members, of negligible stiffness, into the structure to maintain a

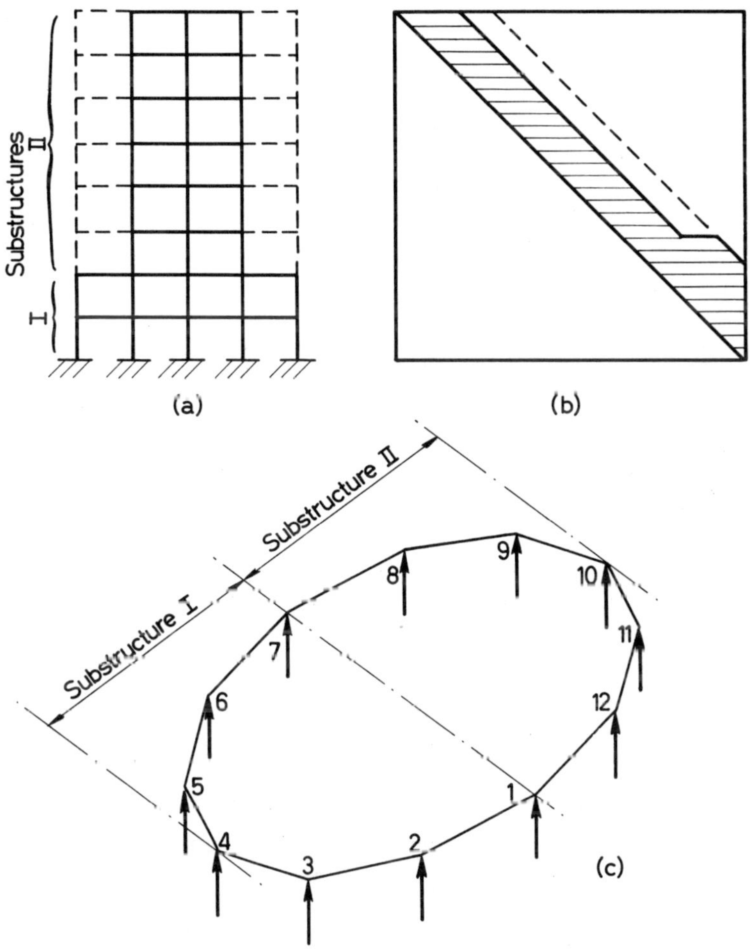

Figure 5.10. Method of substructures

constant band width.[20] Thus, the introduction of the dotted members in Fig. 5.10a will produce a uniform width band in the stiffness matrix, Fig. 5.10b.

Division into substructures may also be useful in ring-like structures, Fig. 5.10c, for which the simple band form is destroyed because of the closed nature of the ring. If the nodes are numbered, as shown in Fig. 5.10c, then the first and last nodes are connected and non-zero cross-stiffnesses will occur remote from the diagonal of the matrix. However, if the structure is divided into substructures I and II, as shown, the stiffnesses of the individual substructures will reveal the narrow band form.

The concept of substructures also permits the solution of large structures on small computers.

Consider a structure divided into substructures I and II. The stiffness matrices of the substructures are arranged in the form

$$\left[\begin{array}{c|c} \mathbf{K}_1 & \mathbf{C}_1^T \\ \hline \mathbf{C}_1 & \mathbf{K}_{1C} \end{array}\right] \left[\begin{array}{c} \mathbf{r}_1 \\ \hline \mathbf{r}_C \end{array}\right] = \left[\begin{array}{c} \mathbf{R}_1 \\ \hline \mathbf{R}_{1C} \end{array}\right] \tag{5.48}$$

$$\left[\begin{array}{c|c} \mathbf{K}_{2C} & \mathbf{C}_2 \\ \hline \mathbf{C}_2^T & \mathbf{K}_2 \end{array}\right] \left[\begin{array}{c} \mathbf{r}_C \\ \hline \mathbf{r}_2 \end{array}\right] = \left[\begin{array}{c} \mathbf{R}_{2C} \\ \hline \mathbf{R}_2 \end{array}\right] \tag{5.49}$$

respectively for substructures I and II. The subscript C is used for the forces and displacements at the nodes common to both substructures. These equations cannot be solved directly since the forces $\mathbf{R}_{1C}$ and $\mathbf{R}_{2C}$ are internal forces exerted by one substructure on the other, and these are unknown. However, we do know that $\mathbf{R}_{1C} = -\mathbf{R}_{2C} = \mathbf{R}_C$ (say). The condensation treatment of section 5.6 may be applied to eqns. (5.48) and (5.49) to give

$$\overline{\mathbf{K}}_{\mathrm{I}C}\mathbf{r}_C = \overline{\mathbf{R}}_{\mathrm{I}C} \tag{5.50}$$

where $\overline{\mathbf{K}}_{\mathrm{I}C} = (\mathbf{K}_{1C} - \mathbf{C}_1\mathbf{K}_1^{-1}\mathbf{C}_1^T)$

and $\overline{\mathbf{R}}_{\mathrm{I}C} = (\mathbf{R}_C - \mathbf{C}_1\mathbf{K}_1^{-1}\mathbf{R}_1)$

and
$$\overline{\mathbf{K}}_{\mathrm{II}C}\mathbf{r}_C = \overline{\mathbf{R}}_{\mathrm{II}C} \tag{5.51}$$

where $\overline{\mathbf{K}}_{\mathrm{II}C} = (\mathbf{K}_{2C} - \mathbf{C}_2\mathbf{K}_2^{-1}\mathbf{C}_2^T)$

and $\overline{\mathbf{R}}_{\mathrm{II}C} = (-\mathbf{R}_C - \mathbf{C}_2\mathbf{K}_2^{-1}\mathbf{R}_2)$

Equations (5.50) and (5.51) may be solved for the internal forces $\mathbf{R}_C$ and displacements $\mathbf{r}_C$ at the common nodes, thus

$$[\overline{\mathbf{K}}_{\mathrm{I}C} + \overline{\mathbf{K}}_{\mathrm{II}C}]\mathbf{r}_C = -[\mathbf{C}_1\mathbf{K}_1^{-1}\mathbf{R}_1 + \mathbf{C}_2\mathbf{K}_2^{-1}\mathbf{R}_2] \tag{5.52}$$

$$[\overline{\mathbf{K}}_{\mathrm{I}C}^{-1} + \overline{\mathbf{K}}_{\mathrm{II}C}^{-1}]\mathbf{R}_C = \overline{\mathbf{K}}_{\mathrm{I}C}^{-1}(\mathbf{C}_1\mathbf{K}_1^{-1}\mathbf{R}_1) - \overline{\mathbf{K}}_{\mathrm{II}C}^{-1}(\mathbf{C}_2\mathbf{K}_2^{-1}\mathbf{R}_2) \tag{5.53}$$

The displacements $\mathbf{r}_1$ may then be obtained from eq. (5.48)

$$\mathbf{r}_1 = \mathbf{K}_1^{-1}[\mathbf{R}_1 - \mathbf{C}_1^T\mathbf{r}_C] \tag{5.54}$$

and the displacements $\mathbf{r}_2$ may be obtained from eq. (5.49)

$$\mathbf{r}_2 = \mathbf{K}_2^{-1}[\mathbf{R}_2 - \mathbf{C}_2^T\mathbf{r}_C] \tag{5.55}$$

PROBLEMS

5.1. Compute the stiffness matrix for the beam of problem 3.1.
5.2. Compute the stiffness matrix for the beam of problem 3.2.
5.3. Compute the stiffnesses and fixed-end moments for the rectangular-section beam shown in Fig. 5.11. $E/G = 2{\cdot}5$.

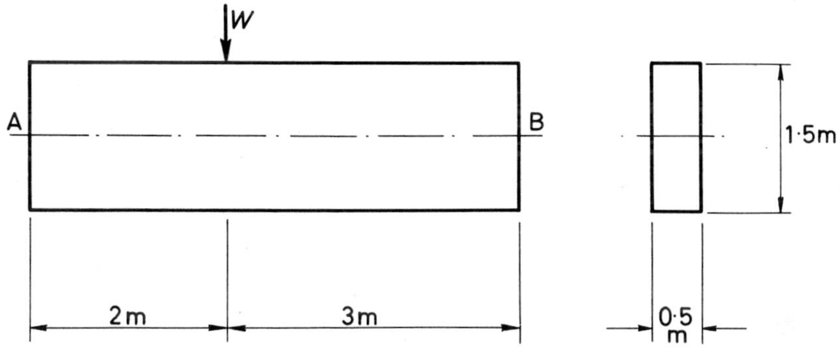

Figure 5.11. Problem 5.3

5.4. Assemble a stiffness matrix for the beam shown in Fig. 5.12, treating the beam as composed of three segments with displacements numbered 1, 2, . . ., 6. Partition the matrix in a form suitable for condensation to produce the end stiffnesses of the beam.

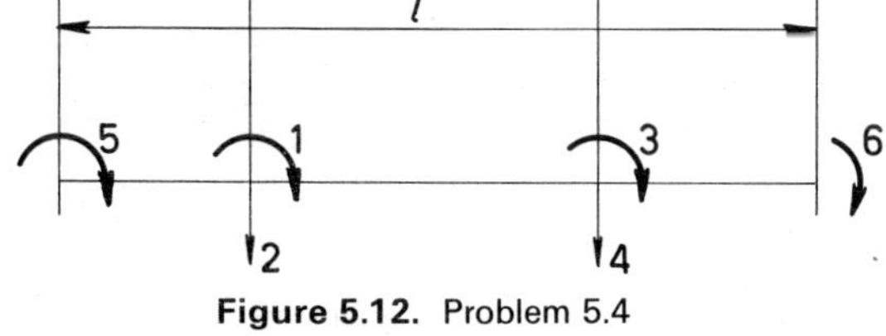

Figure 5.12. Problem 5.4

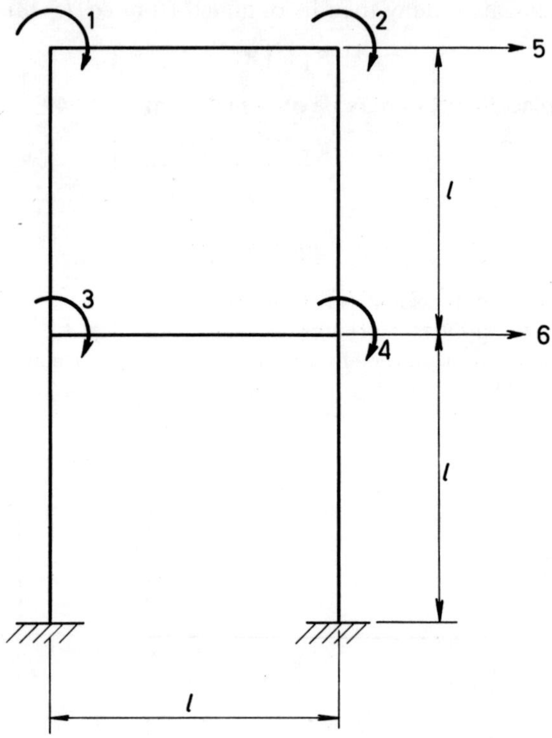

Figure 5.13. Problem 5.5

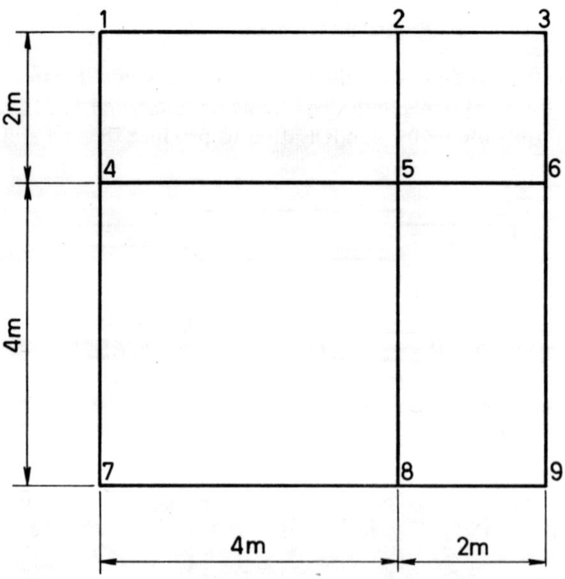

Figure 5.14. Problem 5.6

5.5. Assemble a stiffness matrix for the structure shown in Fig. 5.13, neglecting axial strains in members. Use the method of matrix condensation to obtain the stiffness matrix related to the lateral displacements of the storeys. All members have the same, uniform, EI.

5.6. The grillage shown in Fig. 5.14 is constructed of uniform members having $E = 207\ \text{kN/mm}^2$, $I = 1870 \times 10^3\ \text{mm}^4$, $J = 1960 \times 10^3\ \text{mm}^4$, and $E/G = 2{\cdot}6$. All the members are rigidly connected together at the joints and the grillage is supported with purely vertical restraint at the joints 1, 3, 7, and 9. Determine the nodal displacements and member forces due to a unit transverse load at joint 5.

6. The Finite Element Method

Structural analysis by the finite element technique may be divided into three stages. First, a mathematical model of the structure is formulated in which the structure is represented as an assemblage of discrete parts, called 'elements'. Each element has finite dimensions and properties and, in order to perform the subsequent analysis, it is necessary to establish the force-displacement relationships of each element. The determination of these relationships forms the second stage of the process.

If a pattern of finite elements is used to represent an elastic continuum, such as a plate, the division of the structure can be carried out in many different ways. The elements may, for example, be taken as triangular or quadrilateral plate elements. The representation may be *coarse*, with a small number of elements; or *fine*, using a relatively large number of elements. The actual choice will depend on the geometry of the structure and on the importance of local features such as stress concentrations. It may be desirable to *grade* the pattern of elements so that regions in which stress concentrations are expected, are covered with a finer division into elements than other regions in the structure.

In the case of a framed structure, an obvious and convenient division into elements is obtained if we consider each member, beam or column, as a finite element. As these members are represented mathematically in analysis by an *elastic line*, their connection can only occur at their ends, that is at the joints or *nodes*. Thus, much of the structural analysis we have considered in other chapters comes strictly under the heading of 'finite element' analysis. The main difference is that in framed structures the elements are connected only at the nodes and considerations of equilibrium and displacement compatibility apply only at the nodes. With the types of element normally coming under the title 'finite elements', we are attempting to analyse structures which are represented mathematically by two-dimensional elastic sheets, plates and shells, or by a three-dimensional elastic continuum. In these cases the connection between elements occurs along lines or over faces, and not only at nodal points. The procedure is to ensure equilibrium and compatibility at the connecting nodes of the elements, and constrain the regions of connection between nodes to adopt a chosen form of displacement distribution. Depending on the particular displacement distribution adopted, the compatibility conditions applying along the interfaces of adjacent elements may or may not be satisfied. In cases where compatibility conditions are violated, a degree of approximation is introduced and an assessment of this is required. This aspect of finite element work will be further examined in section 6.7.

The third stage of the process is the analysis of the structure composed of a number, possibly a very large number, of elements, all with known structural properties, and connected together in a prescribed way. This is essentially the process we have already considered in earlier chapters and the methods already developed are available. In general, the stiffness method is found to be the more suitable in finite element work and this method will be adopted in what follows.

This chapter should be regarded as an introduction only to this powerful and rapidly developing method. For a more complete study of the method the reader is referred to the first textbook devoted entirely to the subject[42] and to the growing list of publications appearing in the technical journals.

6.1. Evaluation of Element Stiffnesses—Beam Element

The second stage of the three-stage process mentioned in the introduction is the one requiring most attention. The first stage, that of choosing the pattern of elements to represent a given structure, is not usually difficult, and the last stage of analysis is done by the computer using existing, or easily prepared programs. The process of establishing force-displacement relationships for a given finite element will be introduced with reference to a simple example. Since the stiffness matrix, relating nodal forces and nodal displacements, of the beam element is already well known, eq. (1.27), a study of this simple, one-dimensional, element will suffice to establish the general procedure and enable us to gain confidence in the method.

Consider the uniform beam ij shown in Fig. 6.1. Neglecting any axial strains in the beam, the deformations may be described in terms of the slope

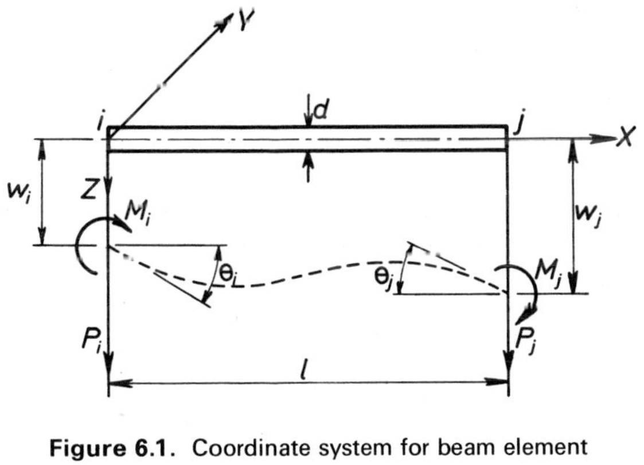

Figure 6.1. Coordinate system for beam element

$\mathrm{d}w/\mathrm{d}x$ and the deflection w at each end of the beam. The deflected shape of the beam may be taken as a polynomial, eq. (6.1), with four arbitrary constants, $\alpha_{1\ldots4}$, corresponding to the four nodal degrees of freedom of the element.

$$w = \alpha_1 + \alpha_2 x + \alpha_3 x^2 + \alpha_4 x^3 \tag{6.1}$$

$$= [1 \quad x \quad x^2 \quad x^3]\{\alpha_1 \quad \alpha_2 \quad \alpha_3 \quad \alpha_4\}$$

$$w = \mathbf{L}\boldsymbol{\alpha} \tag{6.2}$$

Thus

$$\frac{\mathrm{d}w}{\mathrm{d}x} = \alpha_2 + 2\alpha_3 x + 3\alpha_4 x^2$$

$$\frac{\mathrm{d}^2 w}{\mathrm{d}x^2} = 2\alpha_3 + 6\alpha_4 x$$

$$\frac{\mathrm{d}^3 w}{\mathrm{d}x^3} = 6\alpha_4$$

$$\frac{\mathrm{d}^4 w}{\mathrm{d}x^4} = 0$$

Now

$$\theta_i = \left(\frac{\mathrm{d}w}{\mathrm{d}x}\right)_i \quad \text{and} \quad \theta_j = \left(\frac{\mathrm{d}w}{\mathrm{d}x}\right)_j, \quad x_i = 0,\ x_j = l$$

Thus

$$\begin{bmatrix} w_i \\ \theta_i \\ w_j \\ \theta_j \end{bmatrix} = \begin{bmatrix} 1 & 0 & 0 & 0 \\ 0 & 1 & 0 & 0 \\ 1 & l & l^2 & l^3 \\ 0 & 1 & 2l & 3l^2 \end{bmatrix} \begin{bmatrix} \alpha_1 \\ \alpha_2 \\ \alpha_3 \\ \alpha_4 \end{bmatrix}$$

or

$$\boldsymbol{\Delta} = \mathbf{A}\boldsymbol{\alpha} \tag{6.3}$$

The displacement at any point in the element, w, may be expressed in terms of the four nodal displacements $\boldsymbol{\Delta}$ by combining eqns. (6.2) and (6.3), thus

$$w = \mathbf{L}\mathbf{A}^{-1}\boldsymbol{\Delta}$$

$$= \mathbf{C}\boldsymbol{\Delta} \tag{6.4}$$

In accordance with normal beam theory the distribution of strain, and stress, through the depth, d, of the element is linear and thus the state of strain in the element may be represented by the curvature $\mathrm{d}^2 w/\mathrm{d}x^2$.

From eq. (6.1)

$$\mathrm{d}^2 w/\mathrm{d}x^2 = [0 \quad 0 \quad 2 \quad 6x]\{\alpha_1 \quad \alpha_2 \quad \alpha_3 \quad \alpha_4\}$$

or

$$\boldsymbol{\varepsilon} = \mathbf{B}\boldsymbol{\alpha} \tag{6.5}$$

$$= \mathbf{B}\mathbf{A}^{-1}\boldsymbol{\Delta} \quad \text{from eq. (6.3)}$$

The state of 'stress' in the element is defined by the bending moment distribution. Thus, on the understanding that the matrix $\boldsymbol{\sigma}$ defines the state of stress within the element, without necessarily containing the actual values of the stresses, we may put

$$\begin{aligned}\boldsymbol{\sigma} &= EI\,\mathrm{d}^2w/\mathrm{d}x^2\\ &= \mathbf{D}\boldsymbol{\varepsilon} \qquad (6.6)\\ &= \mathbf{DBA}^{-1}\boldsymbol{\Delta}\end{aligned}$$

The matrix **D** is the 'elasticity' matrix relating 'stress' and 'strain', and in this case **D** consists of a single item, namely the flexural rigidity of the beam.

To obtain the stiffnesses of the element we employ the principle of virtual work, applying arbitrary nodal displacements $\bar{\boldsymbol{\Delta}}$, and equating external and internal work.

The virtual strains produced in the element are, from eq. (6.5),

$$\bar{\boldsymbol{\varepsilon}} = \mathbf{BA}^{-1}\bar{\boldsymbol{\Delta}} \qquad (6.7)$$

The internal work in an elemental length of beam dx is

$$\begin{aligned}\bar{\boldsymbol{\varepsilon}}^T\boldsymbol{\sigma}\,\mathrm{d}x &= [\mathbf{BA}^{-1}\bar{\boldsymbol{\Delta}}]^T[\mathbf{DBA}^{-1}]\boldsymbol{\Delta}\,\mathrm{d}x\\ &= \bar{\boldsymbol{\Delta}}^T[\mathbf{A}^{-1}]^T\mathbf{B}^T\mathbf{DBA}^{-1}\boldsymbol{\Delta}\,\mathrm{d}x \qquad (6.8)\end{aligned}$$

using eqns. (6.6) and (6.7).

Noting that the matrix **B** alone is a function of x, the total internal work is

$$\bar{\boldsymbol{\Delta}}^T[\mathbf{A}^{-1}]^T\left\{\int_0^l \mathbf{B}^T\mathbf{DB}\,\mathrm{d}x\right\}[\mathbf{A}^{-1}]\boldsymbol{\Delta}$$

The external work is the product of the virtual displacements $\bar{\boldsymbol{\Delta}}$ and the nodal forces, **S**, thus

$$\bar{\boldsymbol{\Delta}}^T\mathbf{S} = \bar{\boldsymbol{\Delta}}^T[\mathbf{A}^{-1}]^T\left\{\int_0^l \mathbf{B}^T\mathbf{DB}\,\mathrm{d}x\right\}[\mathbf{A}^{-1}]\boldsymbol{\Delta} \qquad (6.9)$$

The virtual displacements are quite arbitrary and in particular may be taken to be represented by a unit matrix, thus eq. (6.9) can be written

$$\mathbf{S} = [\mathbf{A}^{-1}]^T\left\{\int_0^l \mathbf{B}^T\mathbf{DB}\,\mathrm{d}x\right\}[\mathbf{A}^{-1}]\boldsymbol{\Delta} \qquad (6.10)$$

which is the force-displacement relationship for the beam.

Thus the beam stiffness matrix is given by

$$\mathbf{k} = [\mathbf{A}^{-1}]^T\left\{\int_0^l \mathbf{B}^T\mathbf{DB}\,\mathrm{d}x\right\}[\mathbf{A}^{-1}] \qquad (6.11)$$

Now the matrix **A** is given in eq. (6.3) and its inverse is

$$\mathbf{A}^{-1} = \begin{bmatrix} 1 & 0 & 0 & 0 \\ 0 & 1 & 0 & 0 \\ -3/l^2 & -2/l & 3/l^2 & -1/l \\ 2/l^3 & 1/l^2 & -2/l^3 & 1/l^2 \end{bmatrix}$$

From eq. (6.5),

$$\mathbf{B}^T\mathbf{D}\mathbf{B} = EI\begin{bmatrix} 0 & 0 & 0 & 0 \\ 0 & 0 & 0 & 0 \\ 0 & 0 & 4 & 12x \\ 0 & 0 & 12x & 36x^2 \end{bmatrix}$$

Thus

$$\int_0^l \mathbf{B}^T\mathbf{D}\mathbf{B}\,\mathrm{d}x = EI\begin{bmatrix} 0 & 0 & 0 & 0 \\ 0 & 0 & 0 & 0 \\ 0 & 0 & 4l & 6l^2 \\ 0 & 0 & 6l^2 & 12l^3 \end{bmatrix}$$

Hence

$$\mathbf{k} = \frac{EI}{l^3}\begin{bmatrix} 12 & 6l & -12 & 6l \\ 6l & 4l^2 & -6l & 2l^2 \\ -12 & -6l & 12 & -6l \\ 6l & 2l^2 & -6l & 4l^2 \end{bmatrix} \qquad (6.12)$$

which agrees with eq. (1.27).

Uniformly distributed loads. The general expression for the nodal forces has the form

$$\mathbf{S} = \mathbf{k}\boldsymbol{\Delta} + \mathbf{S}_0 \qquad (6.13)$$

where the nodal forces $\mathbf{S}_0$ are those required to hold a distributed load p in equilibrium when $\boldsymbol{\Delta} = \mathbf{0}$. The virtual work principle may be applied again to determine $\mathbf{S}_0$.

For a set of virtual nodal displacements $\bar{\boldsymbol{\Delta}}$, the work done by the equivalent nodal forces is

$$\bar{\boldsymbol{\Delta}}^T\mathbf{S}_0$$

The work done by the distributed load is

$$\int_0^l \mathbf{w}^T p\,\mathrm{d}x$$

and from eq. (6.4) $\qquad \mathbf{w} = \mathbf{L}\mathbf{A}^{-1}\bar{\boldsymbol{\Delta}}$

Equating work done by the equivalent nodal forces to that done by the load,

$$\bar{\Delta}^T \mathbf{S}_0 = -\bar{\Delta}^T [\mathbf{A}^{-1}]^T \int_0^l \mathbf{L}^T p \, dx$$

Thus

$$\mathbf{S}_0 = -[\mathbf{A}^{-1}]^T \int_0^l \mathbf{L}^T p \, dx$$

or

$$\mathbf{S}_0 = -\int_0^l \mathbf{C}^T p \, dx \tag{6.14}$$

Assuming the load to be uniformly distributed over the length of the beam,

$$\begin{aligned} \mathbf{C} &= [1 \quad x \quad x^2 \quad x^3]\mathbf{A}^{-1} \\ &= 1/l^3[l^3 - 3x^2 l + 2x^3,\ l^3 x - 2x^2 l^2 + x^3 l,\ 3x^2 l - 2x^3,\ -x^2 l^2 + x^3 l] \end{aligned}$$

hence

$$\mathbf{S}_0 = \frac{pl}{12}\{-6, \ -l, \ -6, \ l\}$$

which agrees with the well known values for the 'fixed-end' effects of a uniformly distributed load on a uniform beam.

Having established the method of obtaining element force-displacement relationships in the form of a stiffness matrix, we may now go on to consider other types of structural element. It will be convenient to summarize the results so far obtained, and these are set out in Table 6.1.

Table 6.1. Summary of Procedure, Finite Element Method

Eq. 6.1	Displacement function parameters	$\boldsymbol{\alpha}$
6.3	Nodal displacements	$\Delta = \mathbf{A}\boldsymbol{\alpha}$
6.2	Internal displacements	$w = \mathbf{L}\boldsymbol{\alpha}$ $= \mathbf{L}\mathbf{A}^{-1}\Delta$
6.4		$= \mathbf{C}\Delta$
6.5	Internal strains	$\varepsilon = \mathbf{B}\boldsymbol{\alpha} = \mathbf{B}\mathbf{A}^{-1}\Delta$
6.6	Internal stresses	$\sigma = \mathbf{D}\varepsilon = \mathbf{D}\mathbf{B}\mathbf{A}^{-1}\Delta$
6.13	Nodal forces	$\mathbf{S} = \mathbf{k}\Delta + \mathbf{S}_0$
6.11	Element stiffness	$\mathbf{k} = [\mathbf{A}^{-1}]^T \int_{\text{vol}} \mathbf{B}^T\mathbf{D}\mathbf{B} \, d \text{ vol } [\mathbf{A}^{-1}]$
6.14	Nodal (constrained) forces due to distributed load p	$\mathbf{S}_0 = -\int \mathbf{C}^T p \, d \text{ vol}$

6.2. Triangular Element for Plane Stress

The triangular element has obvious advantages since it is possible to represent irregular shapes and boundaries very simply with a division of the area into triangles.

The internal displacement of any point within the triangular element shown in Fig. 6.2 is uniquely defined by the displacements u and v. Thus the nodal displacements may be represented by

$$\Delta = \{u_i \quad u_j \quad u_k \quad v_i \quad v_j \quad v_k\} \tag{6.15}$$

The displacements u, v are taken to be linear functions of x, y as follows:

$$\left.\begin{aligned} u &= \alpha_1 + \alpha_2 x + \alpha_3 y \\ v &= \alpha_4 + \alpha_5 x + \alpha_6 y \end{aligned}\right\} \tag{6.16}$$

Following the scheme summarized in Table 6.1, we express the nodal displacements Δ in terms of the displacement parameters $\boldsymbol{\alpha}$, as follows:

$$\begin{bmatrix} u_i \\ u_j \\ u_k \\ v_i \\ v_j \\ v_k \end{bmatrix} = \begin{bmatrix} 1 & 0 & 0 & 0 & 0 & 0 \\ 1 & a & 0 & 0 & 0 & 0 \\ 1 & c & b & 0 & 0 & 0 \\ 0 & 0 & 0 & 1 & 0 & 0 \\ 0 & 0 & 0 & 1 & a & 0 \\ 0 & 0 & 0 & 1 & c & b \end{bmatrix} \begin{bmatrix} \alpha_1 \\ \alpha_2 \\ \alpha_3 \\ \alpha_4 \\ \alpha_5 \\ \alpha_6 \end{bmatrix} \tag{6.17}$$

that is, $\Delta = \mathbf{A}\boldsymbol{\alpha}$

The strains in the element are

$$\boldsymbol{\varepsilon} = \begin{bmatrix} \varepsilon_x \\ \varepsilon_y \\ \gamma_{xy} \end{bmatrix} = \begin{bmatrix} \partial u/\partial x \\ \partial v/\partial y \\ \partial u/\partial y + \partial v/\partial x \end{bmatrix}$$

$$= \begin{bmatrix} 0 & 1 & 0 & 0 & 0 & 0 \\ 0 & 0 & 0 & 0 & 0 & 1 \\ 0 & 0 & 1 & 0 & 1 & 0 \end{bmatrix} \begin{bmatrix} \alpha_1 \\ \alpha_2 \\ \alpha_3 \\ \alpha_4 \\ \alpha_5 \\ \alpha_6 \end{bmatrix} \tag{6.18}$$

that is, $\boldsymbol{\varepsilon} = \mathbf{B}\boldsymbol{\alpha}$

Since the **B** matrix contains only constant terms, it follows that the strains are constant within the element.

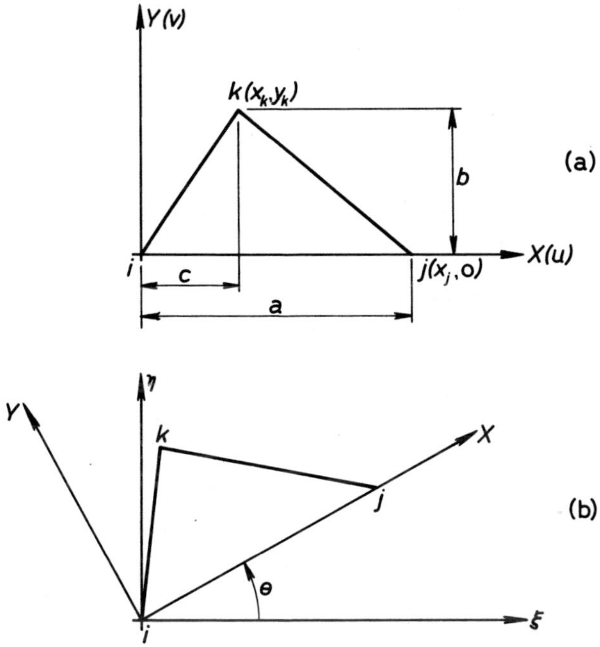

Figure 6.2. Triangular plane stress element

The stress-strain relationships for plane stress in an isotropic material, with Poisson's ratio v, are given by

$$\begin{bmatrix} \sigma_x \\ \sigma_y \\ \tau_{xy} \end{bmatrix} = \frac{E}{1 - v^2} \begin{bmatrix} 1 & v & 0 \\ v & 1 & 0 \\ 0 & 0 & \frac{1}{2}(1 - v) \end{bmatrix} \begin{bmatrix} \varepsilon_x \\ \varepsilon_y \\ \gamma_{xy} \end{bmatrix} \tag{6.19}$$

that is,

$$\boldsymbol{\sigma} = \mathbf{D}\boldsymbol{\varepsilon}$$

We may now obtain the stiffness matrix of the element using step 7 in Table 6.1

$$\mathbf{k} = [\mathbf{A}^{-1}]^T \int_{\text{vol}} \mathbf{B}^T\mathbf{D}\mathbf{B} \, \text{d vol} \, [\mathbf{A}^{-1}]$$

Since the matrices **B** and **D** contain only constant terms, the expression for the stiffness matrix becomes

$$\mathbf{k} = [\mathbf{A}^{-1}]^T\mathbf{B}^T\mathbf{D}\mathbf{B}(V)[\mathbf{A}^{-1}]$$

where V = volume of triangular element

$$= \frac{t}{2}(x_j y_k) = \frac{tab}{2} \tag{6.20}$$

In forming $\mathbf{A}^{-1}$ it is helpful to observe the form of the matrix $\mathbf{A}$ in eq. (6.17),

$$\mathbf{A} = \begin{bmatrix} \mathbf{A}_3 & \mathbf{0}_3 \\ \mathbf{0}_3 & \mathbf{A}_3 \end{bmatrix}$$

where

$$\mathbf{A}_3 = \begin{bmatrix} 1 & 0 & 0 \\ 1 & a & 0 \\ 1 & c & b \end{bmatrix}$$

and $\mathbf{0}_3$ is a (3×3) null matrix.

It is easy to show that

$$\mathbf{A}^{-1} = \begin{bmatrix} \mathbf{A}_3{}^{-1} & \mathbf{0}_3 \\ \mathbf{0}_3 & \mathbf{A}_3{}^{-1} \end{bmatrix} \tag{6.21}$$

where

$$\mathbf{A}_3{}^{-1} = \frac{1}{ab}\begin{bmatrix} ab & 0 & 0 \\ -b & b & 0 \\ (c-a) & -c & a \end{bmatrix} \tag{6.22}$$

Thus, from eq. (6.19), putting $\lambda_1 = \frac{1}{2}(1 - \nu)$ and $\lambda_2 = \frac{1}{2}(1 + \nu)$, we obtain the element stiffness matrix, eq. (6.23).

6.3. Transformation to Global Coordinates

The stiffness matrix (6.23) is based on an X–Y coordinate system in which the X axis is directed along the side ij of the triangular element. The transformation of this stiffness matrix to a global $\xi\eta$ system, as shown in Fig. 6.2b, is carried out as follows:

The relationships between the nodal forces in the global and local coordinate systems is

$$\begin{bmatrix} P_{\xi i} \\ P_{\xi j} \\ P_{\xi k} \\ P_{\eta i} \\ P_{\eta j} \\ P_{\eta k} \end{bmatrix} = \begin{bmatrix} \cos\theta & 0 & 0 & -\sin\theta & 0 & 0 \\ 0 & \cos\theta & 0 & 0 & -\sin\theta & 0 \\ 0 & 0 & \cos\theta & 0 & 0 & -\sin\theta \\ \sin\theta & 0 & 0 & \cos\theta & 0 & 0 \\ 0 & \sin\theta & 0 & 0 & \cos\theta & 0 \\ 0 & 0 & \sin\theta & 0 & 0 & \cos\theta \end{bmatrix} \begin{bmatrix} P_{xi} \\ P_{xj} \\ P_{xk} \\ P_{yi} \\ P_{yj} \\ P_{yk} \end{bmatrix}$$

that is,

$$\bar{\mathbf{S}} = \boldsymbol{\lambda}\mathbf{S} \tag{6.24}$$

$$
\mathbf{k} = \frac{Et}{2(1 - \nu^2)ab}
\begin{array}{c|c|c|c|c|c|c|}
1 & b^2 + \lambda_1(c-a)^2 & \multicolumn{5}{c}{} \\
2 & -b^2 - \lambda_1 c(c-a) & b^2 + \lambda_1 c^2 & \multicolumn{4}{c}{\text{Symmetric}} \\
3 & \lambda_1 a(c-a) & -\lambda_1 ac & \lambda_1 a^2 & \multicolumn{3}{c}{} \\
4 & -\lambda_2 b(c-a) & \lambda_1 cb + \nu b(c-a) & -\lambda_1 ab & \lambda_1 b^2 + (c-a)^2 & \multicolumn{2}{c}{} \\
5 & \lambda_1 b(c-a) + \nu cb & -\lambda_2 bc & \lambda_1 ab & -\lambda_1 b^2 - c(c-a) & \lambda_1 b^2 + c^2 & \\
6 & -\nu ac & \nu ab & 0 & a(c-a) & -ac & a^2 \\
\multicolumn{1}{c}{} & \multicolumn{1}{c}{1} & \multicolumn{1}{c}{2} & \multicolumn{1}{c}{3} & \multicolumn{1}{c}{4} & \multicolumn{1}{c}{5} & \multicolumn{1}{c}{6}
\end{array}
\qquad (6.23)
$$

and following from eq. (5.31), the element stiffness matrix in global coordinates $\xi\eta$ is obtained from

$$\bar{\mathbf{k}} = \boldsymbol{\lambda}\mathbf{k}\boldsymbol{\lambda}^T \tag{6.25}$$

The stresses within the element are given by eq. (6.6),

$$\boldsymbol{\sigma} = \mathbf{DBA}^{-1}\boldsymbol{\Delta}$$

In this case, the product $\mathbf{DBA}^{-1}$ is independent of the x, y coordinates and thus the stresses σ_x, σ_y, and τ_{xy} are constant within the element. The normal practice is to consider the calculated stresses to act at the centroid of the element. Clearly, the approximation will not be good if steep stress gradients are encountered, unless a very fine division into elements is used. For a more detailed discussion of this topic the reader should refer to Zienkiewicz.[42]

6.4. Rectangular Plane Stress Element

A rectangular plane stress element is useful in certain problems, although its use is obviously more restricted than the triangular element. On the other hand, the fact that the rectangle has four nodes permits the introduction of an extra parameter in the displacement function, and permits a more accurate representation of a varying stress field.

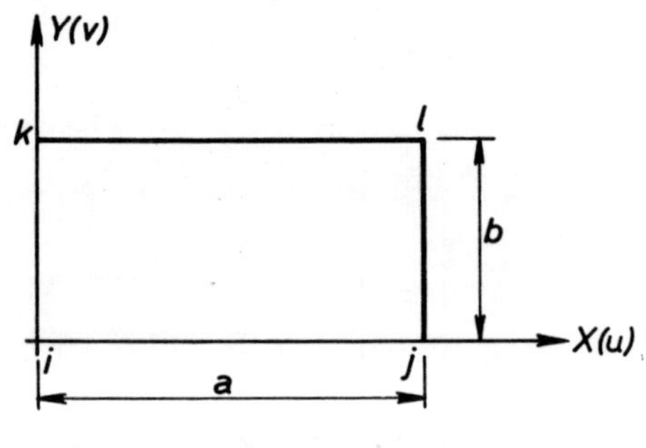

Figure 6.3. Rectangular plane stress element

A rectangular element having dimensions a, b is shown in Fig. 6.3. The development of the stiffness matrix for this element follows closely that used in section 6.2 for the triangular plane stress element. The displacement functions are taken to be

$$\left.\begin{aligned} u &= \alpha_1 + \alpha_2 x + \alpha_3 y + \alpha_4 xy \\ v &= \alpha_5 + \alpha_6 x + \alpha_7 y + \alpha_8 xy \end{aligned}\right\} \tag{6.26}$$

Thus

$$\begin{bmatrix} u_i \\ u_j \\ u_k \\ u_l \\ v_i \\ v_j \\ v_k \\ v_l \end{bmatrix} = \begin{bmatrix} 1 & 0 & 0 & 0 & 0 & 0 & 0 & 0 \\ 1 & a & 0 & 0 & 0 & 0 & 0 & 0 \\ 1 & 0 & b & 0 & 0 & 0 & 0 & 0 \\ 1 & a & b & ab & 0 & 0 & 0 & 0 \\ 0 & 0 & 0 & 0 & 1 & 0 & 0 & 0 \\ 0 & 0 & 0 & 0 & 1 & a & 0 & 0 \\ 0 & 0 & 0 & 0 & 1 & 0 & b & 0 \\ 0 & 0 & 0 & 0 & 1 & a & b & ab \end{bmatrix} \begin{bmatrix} \alpha_1 \\ \alpha_2 \\ \alpha_3 \\ \alpha_4 \\ \alpha_5 \\ \alpha_6 \\ \alpha_7 \\ \alpha_8 \end{bmatrix}$$

that is,

$$\boldsymbol{\Delta} = \mathbf{A}\boldsymbol{\alpha} \tag{6.27}$$

The strains in the element are

$$\begin{bmatrix} \varepsilon_x \\ \varepsilon_y \\ \gamma_{xy} \end{bmatrix} = \begin{bmatrix} \partial u/\partial x \\ \partial v/\partial y \\ (\partial u/\partial y) + (\partial v/\partial x) \end{bmatrix} = \begin{bmatrix} 0 & 1 & 0 & y & 0 & 0 & 0 & 0 \\ 0 & 0 & 0 & 0 & 0 & 0 & 1 & x \\ 0 & 0 & 1 & x & 0 & 1 & 0 & y \end{bmatrix} \begin{bmatrix} \alpha_1 \\ \alpha_2 \\ \alpha_3 \\ \alpha_4 \\ \alpha_5 \\ \alpha_6 \\ \alpha_7 \\ \alpha_8 \end{bmatrix}$$

that is,

$$\boldsymbol{\varepsilon} = \mathbf{B}\boldsymbol{\alpha} \tag{6.28}$$

It is clear from the matrix **B** that the strains are now linear functions of x and y.

The stress-strain relationship is unchanged from eq. (6.19)

$$\begin{bmatrix} \sigma_x \\ \sigma_y \\ \tau_{xy} \end{bmatrix} = \frac{E}{(1 - \nu^2)} \begin{bmatrix} 1 & \nu & 0 \\ \nu & 1 & 0 \\ 0 & 0 & \frac{1}{2}(1 - \nu) \end{bmatrix} \begin{bmatrix} \varepsilon_x \\ \varepsilon_y \\ \gamma_{xy} \end{bmatrix}$$

that is,

$$\boldsymbol{\sigma} = \mathbf{D}\boldsymbol{\varepsilon}$$

The inverse of matrix **A** is found in a similar way to that used in the previous section. Partitioning **A** in the form

$$\mathbf{A} = \begin{bmatrix} \mathbf{A}_4 & \mathbf{0}_4 \\ \mathbf{0}_4 & \mathbf{A}_4 \end{bmatrix}$$

then

$$\mathbf{A}^{-1} = \begin{bmatrix} \mathbf{A}_4{}^{-1} & \mathbf{0}_4 \\ \mathbf{0}_4 & \mathbf{A}_4{}^{-1} \end{bmatrix}$$

where

$$\mathbf{A}_4{}^{-1} = \frac{1}{ab}\begin{bmatrix} ab & 0 & 0 & 0 \\ -b & b & 0 & 0 \\ -a & 0 & a & 0 \\ 1 & -1 & -1 & 1 \end{bmatrix}$$

The element stiffness matrix is now obtained from eq. (6.11)

$$\mathbf{k} = [\mathbf{A}^{-1}]^T \int_{\text{vol}} \mathbf{B}^T\mathbf{D}\mathbf{B}\, \mathrm{d\,vol}\, [\mathbf{A}^{-1}]$$

Now,

$$\mathbf{B}^T\mathbf{D}\mathbf{B} = \frac{Et}{(1-\nu^2)}\begin{bmatrix} 0 & 0 & 0 & 0 & 0 & 0 & 0 & 0 \\ 0 & 1 & 0 & y & 0 & 0 & \nu & \nu x \\ 0 & 0 & \lambda_1 & \lambda_1 x & 0 & \lambda_1 & 0 & \lambda_1 y \\ 0 & y & \lambda_1 x & y^2 + \lambda_1 x^2 & 0 & \lambda_1 x & \nu y & \lambda_2 xy \\ 0 & 0 & 0 & 0 & 0 & 0 & 0 & 0 \\ 0 & 0 & \lambda_1 & \lambda_1 x & 0 & \lambda_1 & 0 & \lambda_1 y \\ 0 & \nu & 0 & \nu y & 0 & 0 & 1 & x \\ 0 & \nu x & \lambda_1 y & \lambda_2 xy & 0 & \lambda_1 y & x & x^2 + \lambda_1 y^2 \end{bmatrix}$$

in which

$$\tfrac{1}{2}(1-\nu) = \lambda_1$$
$$\tfrac{1}{2}(1+\nu) = \lambda_2$$

Performing the necessary integration and matrix multiplications we obtain the element stiffness matrix in eq. (6.29):

$$\mathbf{k} = \frac{Et}{12ab(1 - \nu^2)} \quad (6.29)$$

	1	2	3	4	5	6	7	8
1	$4(b^2 + \lambda_1 a^2)$					Symmetric		
2	$2(\lambda_1 a^2 - 2b^2)$	$4(b^2 + \lambda_1 a^2)$						
3	$2(b^2 - 2\lambda_1 a^2)$	$-2(b^2 + \lambda_1 a^2)$	$4(b^2 + \lambda_1 a^2)$					
4	$-2(b^2 + \lambda_1 a^2)$	$2(b^2 - 2\lambda_1 a^2)$	$2(\lambda_1 a^2 - 2b^2)$	$4(b^2 + \lambda_1 a^2)$				
5	$3ab\lambda_2$	$3ab(\lambda_1 - \nu)$	$3ab(\nu - \lambda_1)$	$-3ab\lambda_2$	$4(a^2 + \lambda_1 b^2)$			
6	$3ab(\nu - \lambda_1)$	$-3ab\lambda_2$	$3ab\lambda_2$	$3ab(\lambda_1 - \nu)$	$2(a^2 - 2\lambda_1 b^2)$	$4(a^2 + \lambda_1 b^2)$		
7	$3ab(\lambda_1 - \nu)$	$3ab\lambda_2$	$-3ab\lambda_2$	$3ab(\nu - \lambda_1)$	$2(\lambda_1 b^2 - 2a^2)$	$-2(a^2 + \lambda_1 b^2)$	$4(a^2 + \lambda_1 b^2)$	
8	$-3ab\lambda_2$	$3ab(\nu - \lambda_1)$	$3ab(\lambda_1 - \nu)$	$3ab\lambda_2$	$-2(a^2 + \lambda_1 b^2)$	$2(\lambda_1 b^2 - 2a^2)$	$2(a^2 - 2\lambda_1 b^2)$	$4(a^2 + \lambda_1 b^2)$

The stresses within the element are given by eq. (6.6),

$$\begin{aligned}\boldsymbol{\sigma} &= [\mathbf{DBA}^{-1}]\boldsymbol{\Delta}\\ &= [\mathbf{Q}]\boldsymbol{\Delta}\end{aligned} \tag{6.30}$$

where

$$\mathbf{Q}=\frac{E}{ab(1-\nu^2)}\begin{bmatrix} y-b & b-y & -y & y & \nu(x-a) & -\nu x & \nu(a-x) & \nu x\\ \nu(y-b) & \nu(b-y) & -\nu y & \nu y & x-a & -x & a-x & x\\ \lambda_1(x-a) & -\lambda_1 x & \lambda_1(a-x) & \lambda_1 x & \lambda_1(y-b) & \lambda_1(b-y) & -\lambda_1 y & \lambda_1 y \end{bmatrix} \tag{6.31}$$

6.5. Parallelogram Element for Plate Flexure

Problems involving plate bending have received a great deal of attention in the development of the finite element method. Plate flexure elements in the form of rectangles and triangles have been formulated[8, 28, 40, 41, 42], and the general quadrilateral element has been studied by Zienkiewicz.[42] The stiffness matrices associated with these elements are available in the literature and may be incorporated in computer programs as required.

A particular element which is useful in skew bridge analysis is the anisotropic parallelogram element and this has received rather less attention in the literature, although the isotropic case has been studied by Dawe.[9] An outline of the development of this element will be given now as an illustration of the use of a skew coordinate system. It should be noted that the final stiffness matrix of the element is not required outside the computer. Once the algebraic and matrix manipulations required to generate the stiffness matrix are determined, these can be programmed and the computer will generate the various matrices as required. Thus, it is unnecessary to produce an explicit form of the stiffness matrix.

The usual 'thin plate' assumptions of linear variation of stress and strain through the thickness of the plate will be adopted in what follows.

The element and the associated coordinate axes are shown in Fig. 6.4. The rectangular, XY, axes and the skew, $\xi\eta$, axes are related as follows:

$$\left.\begin{aligned} x &= \xi + \eta\cos\phi\\ y &= \eta\sin\phi\\ \text{or}\qquad \xi &= x - y\cot\phi\\ \eta &= y\,\mathrm{cosec}\,\phi \end{aligned}\right\} \tag{6.32}$$

The nodal displacements appropriate to plate bending are

$$\boldsymbol{\Delta} = \{\theta_{\xi1}\ \ \theta_{\eta1}\ \ w_1\ \ \theta_{\xi2}\ \ \theta_{\eta2}\ \ w_2\ \ \theta_{\xi3}\ \ \theta_{\eta3}\ \ w_3\ \ \theta_{\xi4}\ \ \theta_{\eta4}\ \ w_4\} \tag{6.33}$$

The deformed shape of the plate can be expressed by the transverse displacement w of the middle surface.

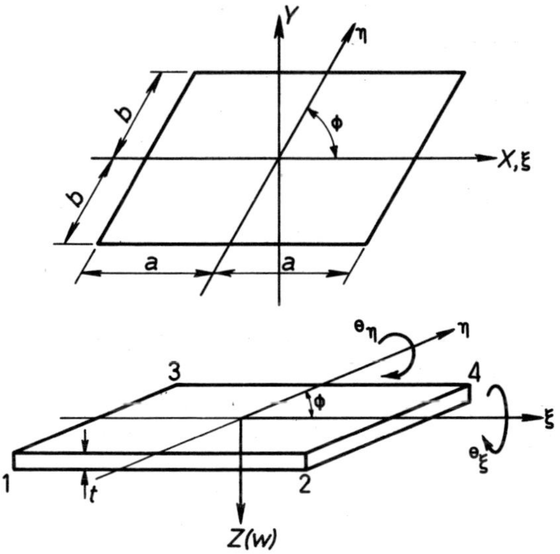

Figure 6.4. Parallelogrammic element for plate flexure

A suitable displacement function containing twelve parameters, corresponding to the twelve nodal displacements, is

$$w = \alpha_1 + \alpha_2\xi + \alpha_3\eta + \alpha_4\xi^2 + \alpha_5\xi\eta + \alpha_6\eta^2 + \alpha_7\xi^3 + \alpha_8\xi^2\eta + \alpha_9\xi\eta^2 + \alpha_{10}\eta^3 + \alpha_{11}\xi^3\eta + \alpha_{12}\xi\eta^3 \quad (6.34)$$

Thus, putting $\theta_\xi = -\partial w/\partial \eta$

and $\theta_\eta = \partial w/\partial \xi$ (6.35)

and substituting nodal coordinates, we obtain the following relationship between the nodal displacements $\boldsymbol{\Delta}$ and the displacement parameters $\boldsymbol{\alpha}$:

$$
\begin{bmatrix} \theta_{\xi 1} \\ \theta_{\eta 1} \\ w_1 \\ \theta_{\xi 2} \\ \theta_{\eta 2} \\ w_2 \\ \theta_{\xi 3} \\ \theta_{\eta 3} \\ w_3 \\ \theta_{\xi 4} \\ \theta_{\eta 4} \\ w_4 \end{bmatrix}
=
\begin{bmatrix}
0 & 0 & -1 & 0 & a & 2b & 0 & -a^2 & -2ab & -3b^2 & a^3 & 3ab^2 \\
0 & 1 & 0 & -2a & -b & 0 & 3a^2 & 2ab & b^2 & 0 & -3a^2b & -b^3 \\
1 & -a & -b & a^2 & ab & b^2 & -a^3 & -a^2b & -ab^2 & -b^3 & a^3b & ab^3 \\
0 & 0 & -1 & 0 & -a & 2b & 0 & -a^2 & 2ab & -3b^2 & -a^3 & -3ab^2 \\
0 & 1 & 0 & 2a & -b & 0 & 3a^2 & -2ab & b^2 & 0 & -3a^2b & -b^3 \\
1 & a & -b & a^2 & -ab & b^2 & a^3 & -a^2b & ab^2 & -b^3 & -a^3b & -ab^3 \\
0 & 0 & -1 & 0 & a & -2b & 0 & -a^2 & 2ab & -3b^2 & a^3 & 3ab^2 \\
0 & 1 & 0 & -2a & b & 0 & 3a^2 & -2ab & b^2 & 0 & 3a^2b & b^3 \\
1 & -a & b & a^2 & -ab & b^2 & -a^3 & a^2b & -ab^2 & b^3 & -a^3b & -ab^3 \\
0 & 0 & -1 & 0 & -a & -2b & 0 & -a^2 & -2ab & -3b^2 & -a^3 & -3ab^2 \\
0 & 1 & 0 & 2a & b & 0 & 3a^2 & 2ab & b^2 & 0 & 3a^2b & b^3 \\
1 & a & b & a^2 & ab & b^2 & a^3 & a^2b & ab^2 & b^3 & a^3b & ab^3
\end{bmatrix}
\begin{bmatrix} \alpha_1 \\ \alpha_2 \\ \alpha_3 \\ \alpha_4 \\ \alpha_5 \\ \alpha_6 \\ \alpha_7 \\ \alpha_8 \\ \alpha_9 \\ \alpha_{10} \\ \alpha_{11} \\ \alpha_{12} \end{bmatrix}
$$

that is,

$$\boldsymbol{\Delta} = \mathbf{A}\boldsymbol{\alpha} \tag{6.36}$$

Since the distribution of stress and strain through the thickness of the plate is assumed linear, the stresses and strains in the element may be defined in terms of the internal stress resultants (M_x, M_y, M_{xy}) and the curvatures

$$(\partial^2 w/\partial x^2, \quad \partial^2 w/\partial y^2, \quad \partial^2 w/\partial x \partial y)$$

The stress-strain relationships take the form[38]

$$\left.\begin{aligned} \sigma_x &= E_x{}'\varepsilon_x + E''\varepsilon_y \\ \sigma_y &= E_y{}'\varepsilon_y + E''\varepsilon_x \\ \tau_{xy} &= G\gamma_{xy} \end{aligned}\right\} \tag{6.37}$$

where

$$\left.\begin{aligned} \varepsilon_x &= -z\,\partial^2 w/\partial x^2 \\ \varepsilon_y &= -z\,\partial^2 w/\partial y^2 \\ \gamma_{xy} &= -2z\,\partial^2 w/\partial x \partial y \end{aligned}\right\} \tag{6.38}$$

Thus

$$\left.\begin{aligned} M_x &= \int_{-t/2}^{+t/2} \sigma_x z\,\mathrm{d}z = -(D_x\,\partial^2 w/\partial x^2 + D_1\,\partial^2 w/\partial y^2) \\ M_y &= \int_{-t/2}^{+t/2} \sigma_y z\,\mathrm{d}z = -(D_y\,\partial^2 w/\partial y^2 + D_1\,\partial^2 w/\partial x^2) \\ M_{xy} &= \int_{-t/2}^{+t/2} \tau_{xy} z\,\mathrm{d}z = 2D_{xy}\,\partial^2 w/\partial x \partial y \end{aligned}\right\} \tag{6.39}$$

where

$$\left.\begin{aligned} D_x &= \frac{E_x{}'t^3}{12} \\ D_y &= \frac{E_y{}'t^3}{12} \\ D_1 &= \frac{E''t^3}{12} \\ D_{xy} &= \frac{Gt^3}{12} \end{aligned}\right\} \tag{6.40}$$

Thus

$$\begin{bmatrix} M_x \\ M_y \\ M_{xy} \end{bmatrix} = \begin{bmatrix} D_x & D_1 & 0 \\ D_1 & D_y & 0 \\ 0 & 0 & 2D_{xy} \end{bmatrix} \begin{bmatrix} -\partial^2 w/\partial x^2 \\ -\partial^2 w/\partial y^2 \\ \partial^2 w/\partial x \partial y \end{bmatrix}$$

or

$$\mathbf{M} = \mathbf{D}\boldsymbol{\varepsilon} \tag{6.41}$$

The curvatures in the skew coordinate system are obtained from those in the rectangular system as follows:

from eqs. (6.32)

$$\partial\xi/\partial x = 1, \quad \partial\xi/\partial y = -\cot\phi, \quad \partial\eta/\partial x = 0, \quad \partial\eta/\partial y = \operatorname{cosec}\phi$$

Now $\partial w/\partial x = \partial w/\partial\xi \; \partial\xi/\partial x + \partial w/\partial\eta \; \partial\eta/\partial x = \partial w/\partial\xi$

thus $\partial/\partial x \equiv \partial/\partial\xi$

and $\partial^2 w/\partial x^2 = \partial^2 w/\partial\xi^2$ (6.42)

Similarly,

$$\begin{aligned}
\partial w/\partial y &= \partial w/\partial\xi \; \partial\xi/\partial y + \partial w/\partial\eta \; \partial\eta/\partial y \\
&= \partial w/\partial\xi \, (-\cot\phi) + \partial w/\partial\eta \, (\operatorname{cosec}\phi) \\
&= (-\cot\phi \; \partial/\partial\xi + \operatorname{cosec}\phi \; \partial/\partial\eta)w
\end{aligned}$$

thus $\partial/\partial y \equiv -\cot\phi \; \partial/\partial\xi + \operatorname{cosec}\phi \; \partial/\partial\eta$

and
$$\begin{aligned}
\partial^2 w/\partial y^2 &= \partial/\partial y \, (\partial w/\partial y) \\
&= \cot^2\phi(\partial^2 w/\partial\xi^2) - 2\cot\phi \operatorname{cosec}\phi(\partial^2 w/\partial\xi\partial\eta) \\
&\qquad + \operatorname{cosec}^2\phi(\partial^2 w/\partial\eta^2)
\end{aligned} \tag{6.43}$$

Further,

$$(\partial^2 w/\partial x\partial y) = -\cot\phi(\partial^2 w/\partial\xi^2) + \operatorname{cosec}\phi(\partial^2 w/\partial\xi\partial\eta) \tag{6.44}$$

Combining eqns. (6.42), (6.43), and (6.44),

$$\begin{bmatrix} -\partial^2 w/\partial x^2 \\ -\partial^2 w/\partial y^2 \\ \partial^2 w/\partial x\partial y \end{bmatrix} = \begin{bmatrix} 1 & 0 & 0 \\ \cot^2\phi & \operatorname{cosec}^2\phi & 2\cot\phi\operatorname{cosec}\phi \\ \cot\phi & 0 & \operatorname{cosec}\phi \end{bmatrix} \begin{bmatrix} -\partial^2 w/\partial\xi^2 \\ -\partial^2 w/\partial\eta^2 \\ \partial^2 w/\partial\xi\partial\eta \end{bmatrix}$$

or
$$\boldsymbol{\varepsilon}_{xy} = \mathbf{H}\boldsymbol{\varepsilon}_{\xi\eta} \tag{6.45}$$

By appropriate differentiations of eq. (6.34),

$$\begin{aligned}
\partial^2 w/\partial\xi^2 &= 2\alpha_4 + 6\alpha_7\xi + 2\alpha_8\eta + 6\alpha_{11}\xi\eta \\
\partial^2 w/\partial\eta^2 &= 2\alpha_6 + 2\alpha_9\xi + 6\alpha_{10}\eta + 6\alpha_{12}\xi\eta \\
\partial^2 w/\partial\xi\partial\eta &= \alpha_5 + 2\alpha_8\xi + 2\alpha_9\eta + 3\alpha_{11}\xi^2 + 3\alpha_{12}\eta^2
\end{aligned}$$

that is,
$$\boldsymbol{\varepsilon}_{\xi\eta} = \mathbf{B}\boldsymbol{\alpha} \tag{6.46}$$

where

$$\mathbf{B} = \begin{bmatrix} 0 & 0 & 0 & -2 & 0 & 0 & -6\xi & -2\eta & 0 & 0 & -6\xi\eta & 0 \\ 0 & 0 & 0 & 0 & 0 & -2 & 0 & 0 & -2\xi & -6\eta & 0 & -6\xi\eta \\ 0 & 0 & 0 & 0 & 1 & 0 & 0 & 2\xi & 2\eta & 0 & 3\xi^2 & 3\eta^2 \end{bmatrix}$$

Following the procedure established in section 6.1, arbitrary nodal displacements $\bar{\Delta}$ are accompanied by virtual strains in the element, $\bar{\varepsilon}$, and the total internal work, W_I, is

$$W_I = \iint \bar{\varepsilon}_{xy}{}^T \sigma_{xy} \, dx \, dy = \iint (\mathbf{H}\bar{\varepsilon}_{\xi\eta})^T (\mathbf{DH}\varepsilon_{\xi\eta}) \, dx \, dy$$

$$= \iint (\bar{\varepsilon}_{\xi\eta})^T \mathbf{H}^T \mathbf{DH} \varepsilon_{\xi\eta} dx \, dy$$

$$= \iint (\mathbf{BA}^{-1}\bar{\Delta})^T \mathbf{H}^T \mathbf{DHBA}^{-1} \Delta \, dx \, dy$$

$$W_I = \bar{\Delta}^T [\mathbf{A}^{-1}]^T \iint \mathbf{B}^T \mathbf{H}^T \mathbf{DHB} \, dx \, dy \, [\mathbf{A}^{-1}] \Delta \qquad (6.47)$$

As before, eq. (6.9), the external work is $\bar{\Delta}^T \mathbf{S}$. Equating internal and external work,

$$\mathbf{S} = [\mathbf{A}^{-1}]^T \iint \mathbf{B}^T \mathbf{H}^T \mathbf{DHB} \, dx \, dy \, [\mathbf{A}^{-1}] \Delta$$

that is,

$$\mathbf{k} = [\mathbf{A}^{-1}]^T \iint \mathbf{B}^T \mathbf{H}^T \mathbf{DHB} \, dx \, dy \, [\mathbf{A}^{-1}] \qquad (6.48)$$

The integrations indicated in eq. (6.48) cannot be carried out directly since the matrix **B** is in terms of $\xi\eta$. A Jacobian[18] is used to transform the integration to the variables $\xi\eta$. If the relations between the two sets of variables are

$$x = \chi(\xi, \eta)$$

and

$$y = \psi(\xi, \eta)$$

then the Jacobian is the determinant,

$$\begin{vmatrix} \partial\chi/\partial\xi & \partial\chi/\partial\eta \\ \partial\psi/\partial\xi & \partial\psi/\partial\eta \end{vmatrix}$$

which, from eqs. (6.32), is

$$\begin{vmatrix} 1 & \cos\phi \\ 0 & \sin\phi \end{vmatrix} = \sin\phi$$

Thus, eq. (6.48) becomes

$$\mathbf{k} = \sin\phi [\mathbf{A}^{-1}]^T \int_{-b}^{b} \int_{-a}^{a} \mathbf{B}^T \mathbf{H}^T \mathbf{DHB} \, d\xi \, d\eta \, [\mathbf{A}^{-1}] \qquad (6.49)$$

Computation of element stiffness matrix. The matrix product $\mathbf{H}^T\mathbf{D}\mathbf{H}$ is easily computed and contains only constant terms. The matrix product $\mathbf{B}^T(\mathbf{H}^T\mathbf{D}\mathbf{H})\mathbf{B}$ may be expressed conveniently in terms of the constant elements of $\mathbf{H}^T\mathbf{D}\mathbf{H}$. The double integration is then performed and the result programmed for computer assembly. The inversion of matrix **A** is carried out by the computer and the final matrix multiplication performed to produce the stiffness matrix. This matrix is not required outside the computer.

Isotropic material. The stiffness matrix, eq. (6.49), can be used for an isotropic element by putting

$$\mathbf{D} = \frac{Et^3}{12(1 - \nu^2)} \begin{bmatrix} 1 & \nu & 0 \\ \nu & 1 & 0 \\ 0 & 0 & \frac{1}{2}(1 - \nu) \end{bmatrix} \tag{6.50}$$

Uniformly distributed loads. The nodal force equivalents, $\mathbf{S}_0$, to a uniformly distributed load p are evaluated in a similar way to that used in section 6.1.

The equivalent nodal forces are obtained from

$$\mathbf{S}_0 = p[\mathbf{A}^{-1}]^T \iint \mathbf{L}^T \, \mathrm{d}x \, \mathrm{d}y$$

and transforming from x, y to ξ, η,

$$\begin{aligned} \mathbf{S}_0 &= p \sin \phi [\mathbf{A}^{-1}]^T \iint \mathbf{L}^T \, \mathrm{d}\xi \, \mathrm{d}\eta \\ &= p \sin \phi [\mathbf{A}^{-1}]^T \left\{ 4ab \quad 0 \quad 0 \quad \frac{4ba^3}{3} \quad 0 \quad \frac{4ab^3}{3} \quad 0 \quad 0 \quad 0 \quad 0 \quad 0 \quad 0 \right\} \end{aligned}$$

The equivalent nodal forces are easily calculated at this stage since the matrix $[\mathbf{A}^{-1}]^T$ is already available in the computer.

Determination of bending and twisting moments. The internal bending and twisting moments are given by eq. (6.41)

$$\mathbf{M} = \mathbf{D}\boldsymbol{\varepsilon}$$

Substituting for $\boldsymbol{\varepsilon}$ from eqns. (6.46) and (6.36)

$$\begin{aligned} \mathbf{M} &= \mathbf{D}\mathbf{B}\mathbf{A}^{-1}\Delta \\ &= [\mathbf{Q}]\Delta \end{aligned}$$

6.6. Assembly of Structure Stiffness Matrix

As an example of the assembly of the stiffness matrix of a complete structure from the stiffness matrices of the individual elements, consider the shear wall shown in Fig. 6.5, in which the wall is represented by a 2 × 4 pattern of rectangular plane stress elements.

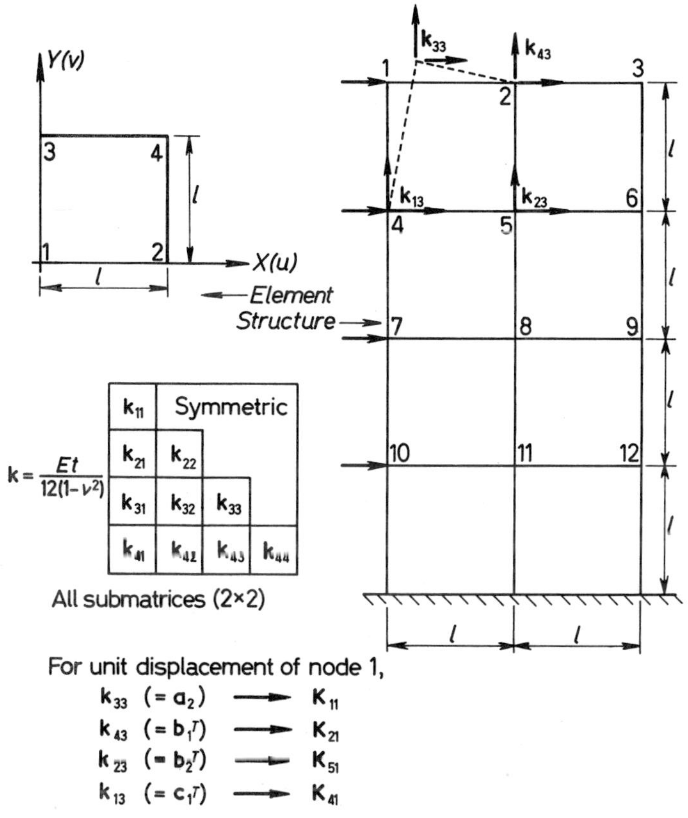

Figure 6.5. Assembly of stiffness matrix **K** for shear wall represented by finite elements

The element stiffness matrix is given by eq. (6.29), and in this case the elements are all identical with $a = b = l$. It is convenient here to rearrange the rows and columns of the element stiffness matrix so that the ordering of the nodal forces and displacements of matrix (6.27) becomes

$$u_i \quad v_i \quad u_j \quad v_j \quad u_k \quad v_k \quad u_l \quad v_l$$

The stiffness matrix of eq. (6.29) then becomes

$$\mathbf{k} = \frac{Et}{12(1 - \nu^2)}$$

	1	2	3	4	5	6	7	8
1	$4(1 + \lambda_1)$					Symmetric		
2	$3\lambda_2$	$4(1 + \lambda_1)$						
3	$2(\lambda_1 - 2)$	$3(\lambda_1 - \nu)$	$4(1 + \lambda_1)$					
4	$3(\nu - \lambda_1)$	2ν	$-3\lambda_2$	$4(1 + \lambda_1)$				
5	2ν	$3(\nu - \lambda_1)$	$-2(1 + \lambda_1)$	$3\lambda_2$	$4(1 + \lambda_1)$			
6	$3(\lambda_1 - \nu)$	$2(\lambda_1 - 2)$	$3\lambda_2$	$-2(1 + \lambda_1)$	$-3\lambda_2$	$4(1 + \lambda_1)$		
7	$-2(1 + \lambda_1)$	$-3\lambda_2$	2ν	$3(\lambda_1 - \nu)$	$2(\lambda_1 - 2)$	$3(\nu - \lambda_1)$	$4(1 + \lambda_1)$	
8	$-3\lambda_2$	$-2(1 + \lambda_1)$	$3(\nu - \lambda_1)$	$2(\lambda_1 - 2)$	$3(\lambda_1 - \nu)$	2ν	$3\lambda_2$	$4(1 + \lambda_1)$

Partitioning the matrix into 2 × 2 submatrices, as shown by the dotted lines, gives

$$\mathbf{k} = \frac{Et}{12(1 - \nu^2)} \begin{bmatrix} \mathbf{a}_1 & & & \text{Symmetric} \\ \mathbf{b}_1 & \mathbf{a}_2 & & \\ \mathbf{c}_1 & \mathbf{b}_2 & \mathbf{a}_2 & \\ \mathbf{d}_1 & \mathbf{c}_1{}^T & \mathbf{b}_1{}^T & \mathbf{a}_1 \end{bmatrix} \tag{6.51}$$

Each of the 2 × 2 submatrices in eq. (6.51) now contains stiffnesses relating to à single node and the stiffness matrix of the wall may be assembled from the 2 × 2 submatrices of the individual elements. The assembly process may be carried out by inspection if we prescribe unit values to each pair of nodal displacements in turn, and enter the appropriate submatrices into the structure stiffness matrix.

The element nodes and structure nodes are numbered as shown in Fig. 6.5. As an illustration of the process of assembling the structure stiffness matrix **K**, consider the generation of the first column of **K**. If we apply unit values of displacements u, v to node 1 of the structure, the only element stiffnesses involved are those pertaining to the element between nodes 1, 2, 4, and 5. The direct stiffness $\mathbf{k}_{33}$ of the element transfers to submatrix $\mathbf{K}_{11}$ in the structure stiffness matrix **K**. The cross-stiffnesses also transfer to submatrix positions in **K**, dependent on the nodal numbers in the structure. The process will be recognized as that used in section 4.8 and called the 'code number' method. Thus, the first column of **K** is formed. The complete matrix is, eq. (6.52).

A regular division of the wall into elements gives rise to a regular pattern in the stiffness matrix, as is evident in eq. (6.52). Where such a regular pattern exists advantage can be taken of it in the programming of the matrix assembly and in the solution of the equations.[22]

The application of the method to the analysis of a perforated shear wall will be considered in detail in section 8.4.

Boundary conditions of support. These were automatically included in the above example, simply by omitting the zero displacements along the bottom edge of the wall. In general, certain displacements may be prescribed at nodal points in the structure and this situation can be handled by planting zeros in the row and column of the stiffness matrix corresponding to the prescribed displacement with a 'one' in the diagonal position. If the corresponding load term in the right-hand side matrix is then replaced by the value of the prescribed displacement, which may be zero, the solution of the equations will give the desired result for that particular displacement.

$$
\mathbf{K} = \frac{Et}{12(1-\nu^2)}
\begin{bmatrix}
\mathbf{a}_2 & & & & & & & & & & & \\
\mathbf{b}_1^T & \mathbf{a}_1+\mathbf{a}_2 & & & & & & \text{Symmetric} & & & & \\
\mathbf{0} & \mathbf{b}_1^T & \mathbf{a}_1 & & & & & & & & & \\
\mathbf{c}_1^T & \mathbf{d}_1 & \mathbf{0} & \mathbf{a}_1+\mathbf{a}_2 & & & & & & & & \\
\mathbf{b}_2 & \mathbf{c}_1+\mathbf{c}_1^T & \mathbf{d}_1 & \mathbf{b}_1+\mathbf{b}_1^T & 2(\mathbf{a}_1+\mathbf{a}_2) & & & & & & & \\
 & \mathbf{b}_2 & \mathbf{c}_1 & \mathbf{0} & \mathbf{b}_1+\mathbf{b}_1^T & \mathbf{a}_1+\mathbf{a}_2 & & & & & & \\
 & & \mathbf{0} & \mathbf{c}_1^T & \mathbf{d}_1 & \mathbf{0} & \mathbf{a}_1+\mathbf{a}_2 & & & & & \\
 & & & \mathbf{b}_2 & \mathbf{c}_1+\mathbf{c}_1^T & \mathbf{d}_1 & \mathbf{b}_1+\mathbf{b}_1^T & 2(\mathbf{a}_1+\mathbf{a}_2) & & & & \\
 & & & & \mathbf{b}_2 & \mathbf{c}_1 & \mathbf{0} & \mathbf{b}_1+\mathbf{b}_1^T & \mathbf{a}_1+\mathbf{a}_2 & & & \\
 & & & & & \mathbf{0} & \mathbf{c}_1^T & \mathbf{d}_1 & \mathbf{0} & \mathbf{a}_1+\mathbf{a}_2 & & \\
 & & & & & & \mathbf{b}_2 & \mathbf{c}_2+\mathbf{c}_1^T & \mathbf{d}_1 & \mathbf{b}_1+\mathbf{b}_1^T & 2(\mathbf{a}_1+\mathbf{a}_2) & \\
 & & & & & & & \mathbf{b}_2 & \mathbf{c}_1 & \mathbf{0} & \mathbf{b}_1+\mathbf{b}_1^T & \mathbf{a}_1+\mathbf{a}_2
\end{bmatrix}
\qquad (6.52)
$$

6.7. Displacement Continuity Between Elements

The formulation of the finite element method ensures compatibility in displacements between adjacent elements *at the nodes*. This follows from the definition of single-valued nodal displacements, and in a structure composed of line elements, beams, columns, etc., no further compatibility conditions are required since the elements are in contact only at the nodes.

With the types of element described in this chapter, the elements are in contact along interfaces joining nodes and it is necessary to examine the requirements of displacement compatibility along these interfaces, and to assess the extent to which these requirements are met by the particular formulation adopted. The displacements occurring within each element are defined by the particular displacement function, or *shape* function, assumed, and it is the form of this function which determines whether compatibility is achieved or not.

In the triangular plane stress element of section 6.2, the element displacements u and v are linear functions of x and y, and thus displacement compatibility is automatically satisfied along the lines joining the nodes, and, by definition, at the nodes. However, strain and stress within the element are constant and thus the accuracy with which a varying stress field can be represented will depend on the fineness of the pattern of division into elements. In regions of relatively large strain gradient, it will be necessary to adopt a finer division than in areas of low strain gradient.

The increased degree of freedom of the rectangular plane stress element, due to the introduction of an extra node, permits the use of a four-term polynomial, eq. (6.26), in place of the three-term polynomial of eq. (6.16). The form of the polynomial used, eq. (6.26), ensures linear variations of displacements along the edges of the elements. Thus, continuity is maintained between adjacent elements. However, the matrix **B**, eq. (6.28), which relates the element strains to the displacement parameters, is now a function of x and y, and it is evident from the form of eq. (6.28) that the strains are linear functions of x, y. The rectangular element can, therefore, be used to represent a varying stress field with a coarser division into elements than would be required using triangular elements. The rectangular element is, however, because of its geometry, more restricted in use than the triangular element. An example of the use of a rectangular plane stress element will be found in section 8.4 and Fig. 8.6, where a graded mesh is used to reveal the stress concentration occurring at the corners of the openings in a perforated shear wall.

With finite elements intended to represent plates under flexure, displacement compatibility between elements must be investigated with regard to the transverse displacement w and its derivatives, constituting the slopes θ_x and θ_y. A displacement function which satisfies these requirements is termed a 'conforming shape function'. Such functions are difficult to find and manipulate in the element derivation, and it is usual practice to adopt a 'non-conforming' shape function. Thus, it is usual to prescribe continuity in the transverse displacement w at element interfaces but violate the requirements of slope continuity. Such approximations will not necessarily lead to inferior accuracy in the actual solution of problems, and in general very acceptable accuracy can be obtained with non-conforming functions. Full compatibility in w, θ_x, and θ_y is of course automatically implied at all the nodes.

For a fuller discussion on this important aspect of finite element work, the reader may consult Zienkiewicz.[42]

7. Digital Computers and Computational Methods for Band Matrices

This chapter is intended to serve as a brief introduction to digital computers and certain computational methods suitable for computers and relevant to structural analysis. The programming languages most frequently used for structural computations are FORTRAN and ALGOL, and the reader wishing to program his own calculations is advised to learn one of these languages. For this purpose, the relevant programming manual will provide all the necessary information. The reader who does not wish to do the actual programming may find this brief introduction of some value in helping him organize his calculations, and communicate with the programmer who is actually writing the program. A great deal of useful information on programming for civil engineering problems is contained in the book by Fenves.[10]

The solution of sets of simultaneous equations is a standard procedure on digital computers and all the available machines have standard programs for carrying out this operation. It is not usually a difficult matter to incorporate these routines in one's own program. However, standard routines for solving sets of equations for which the coefficient matrix is a symmetric band are not so readily available and at the end of this chapter two such programs will be given, one in FORTRAN and one in ALGOL. It is a relatively simple matter to insert these programs as subroutines into a structural analysis program requiring them.

7.1. The Digital Computer

As its name implies, the digital computer calculates with numerical data, as distinct from the analogue computer which operates on the basis of measuring physical quantities such as lengths or voltages. Two features of the electronic digital computer need emphasis from the outset. The computer operates with a *stored program*. The set of instructions which form the program are stored by the machine before they are obeyed. The computer then functions *automatically* in following the instructions one by one. The normal process is to obey the instructions serially, in the order in which they are written, very frequently however an instruction will cause the machine to *jump* to some other part of the program, possibly to repeat a *cycle* of operations. The jump may be *unconditional* or it may be *conditional* on a comparison of magnitude of two numbers, or the sign of a number, for example.

The digital computer has, from the point of view of the user, five basic components. These are shown in block diagram form in Fig. 7.1.

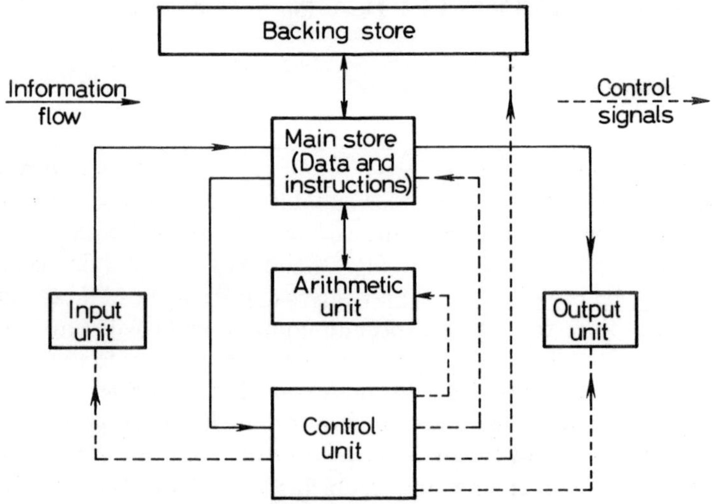

Figure 7.1. Basic components in digital computer

(a) *The storage unit*, or memory, where the instructions and numerical data being operated on, are stored. The instructions, as well as the data, are stored in numerical form using binary digits for representation. This need not present any difficulty to the user who will normally communicate with the machine using decimal digits for his data and a combination of alphanumeric characters for his program. The machine carries out all the necessary conversions. The basic storage unit is usually augmented by a somewhat slower, and less expensive, *backing store*. The backing store may be used to store data and program and is useful in calculations requiring a large provision of storage. Program and data can be moved between the main store and the backing store by instructions in the program. The computer store is divided into *storage registers*, each containing the equivalent of about twelve decimal digits in *floating point* form, i.e., a, 10^b, where a, the argument, contains the significant digits of the number and the exponent, b, is the power of 10 by which the argument has to be multiplied. Thus, all numbers can be stored to the same number of significant digits whether they are large or small. Other storage registers are provided for integer quantities and these can only be used for integer values. Each storage register has an *address*, a number which identifies the register.

(b) *The arithmetic unit* carries out the elementary arithmetical operations of addition, subtraction, multiplication, and division.

(c) *The control unit* provides the facilities for the automatic control of the operations. It selects the instruction currently to be obeyed and serves to control the input and output of data.

(d) *The input unit* serves to enter data and program into the computer store and may consist of punched card or paper tape reader or magnetic tape reader.

(e) *The output unit* may take the form of a card or tape punch, line printer, or graphical plotter.

7.2. The Program

When a computer is used to perform a calculation the entire sequence of operations must be planned out in advance. This list of operations is then translated into a language, or code, acceptable to the machine. Each machine has a number of such codes available to the user who will choose the one most suitable to the job in hand. The complete list of instructions in coded form constitutes the *program*. The program will not be in the basic language of the machine and it will be necessary to process or *compile* the program into the basic machine code. This is done automatically by a special program called a *compiler* which translates the program written by the user into the basic language of the machine. As the program is being compiled the compiler will take notice of any faults in the programmed instructions and indicate the type and locations of the faults; these will be errors in writing down the instructions making them unintelligible to the machine. It will be appreciated that each instruction must be written in a completely unambiguous way and in accordance with the rules laid down for the particular code. The process of testing a program to find out whether an acceptable compiled version can be obtained by the machine is usually termed '*program development*'.

In writing programs it is clearly desirable to economize on the number of instructions as far as possible, since an unnecessarily long program is wasteful of punching time and computer storage. Sequences of operations which are to be repeated for changing values of certain parameters are programmed once and entered whenever required. These *cycles* of operations, together with the great speed of operation, make the computer such a useful computational aid.

Certain standard operations and functions such as square root, trigonometrical and hyperbolic functions, and the operations of matrix algebra are provided for in the form of *subroutines*, which are permanently available to the programmer for incorporation in his program. The programmer will also write his own subroutines for operations which occur in his calculations sufficiently often.

The band matrix solution programs given at the end of this chapter may be treated as subroutines. Subroutines are written with certain parameters, which are only declared when the subroutine is brought into use. For example, two parameters are required to define the size of a band matrix, B the semiband width and N the number of equations. The subroutine which solves the equations is programmed assuming B and N to be variables. Actual numerical values of B and N are placed in specified storage registers just before the subroutine is brought into operation.

The use of *program parameters* permits the same program to be used to solve problems of the same type having different numerical data. The data for such problems will consist of the values of the parameters followed by the other data required, the quantity of this usually being a function of the program parameters.

7.3. Program Organization—The Flow Diagram

The first stage in applying a problem to a computer is to produce a mathematical formulation of the method. This must consist of a series of arithmetical steps and logical decisions. If the programmer is thoroughly familiar with the problem and the proposed method of calculation, he may proceed with the coding operation without giving a great deal of thought to the organization of the calculations. If the programmer is not familiar with the calculation proposed he needs a very clear, unambiguous scheme to guide him in the coding of the problem. A *flow diagram* may be used for this purpose. Consider the process of multiplication of two matrices:

$$\mathbf{AB} = \mathbf{C} \tag{7.1}$$

where the matrices are all of dimensions ($n \times n$). The rule for matrix multiplication may be written as

$$c_{ij} = \sum_{r=1}^{n} a_{ir}b_{rj} \qquad \begin{Bmatrix} i = 1, 2, \ldots, n \\ j = 1, 2, \ldots, n \end{Bmatrix} \tag{7.2}$$

A flow diagram for this operation is shown in Fig. 7.2. The convention adopted here for the block symbols should be noted. The reader is free to devise other conventions he considers suitable for his problem.

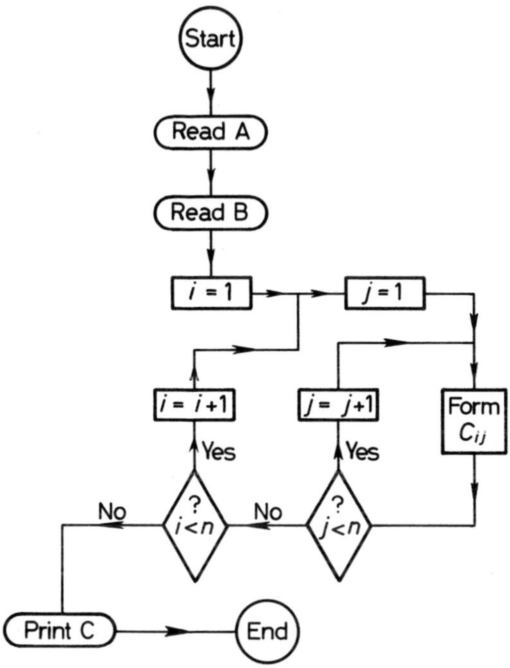

Figure 7.2. Flow diagram for matrix multiplication

7.4. Band Matrix Solution Methods

In previous chapters we have seen that the analysis of a structure involves the solution of a set of linear simultaneous equations. In the flexibility method the equations express the conditions of compatibility in the structure, and in the stiffness method the equations express the conditions of equilibrium relating internal and external forces at the nodes of the structure. The solution of such sets of simultaneous equations is a routine matter on a digital computer which is equipped with standard programs for this purpose.

A very wide range of numerical methods exist for the solution of linear simultaneous equations. The reader is referred to some of the many books on this subject.[43–47] Both direct and iterative methods are available, but, in general, the direct methods are preferred for structural analysis since the programming of these is simpler and they are more easily adapted for the simultaneous analysis of a number of loading systems.

The standard programs will normally assume that the coefficient matrix is fully populated and thus no advantage will be taken of the banded nature of the matrix which frequently occurs in structural problems. Thus, excessive demands may be made on computer storage and time, and on data preparation. In the example cited in section 5.5, Fig. 5.7, the storage required for the symmetric half-band of the stiffness matrix is only $4\frac{1}{2}$ per cent of that required for the complete matrix. Clearly, it is a matter of some importance to have efficient routines for solving matrices of this type and in this section we shall consider a few of the methods on which such routines can be based.

Gauss elimination. In this method, the coefficient matrix is reduced to upper triangular form by successive elimination of the variables. Consider the set of n linear simultaneous equations, eq. (7.3),

$$\begin{bmatrix} a_{11} & a_{12} & \dots & a_{1n} \\ a_{21} & a_{22} & \dots & a_{2n} \\ \vdots & \vdots & & \vdots \\ a_{n1} & a_{n2} & \dots & a_{nn} \end{bmatrix} \begin{bmatrix} x_1 \\ x_2 \\ \vdots \\ x_n \end{bmatrix} = \begin{bmatrix} c_1 \\ c_2 \\ \vdots \\ c_n \end{bmatrix} \tag{7.3}$$

The variable x_1 is eliminated from the last $(n - 1)$ equations by suitable operations with the first equation. The variable x_2 is then eliminated from the last $(n - 2)$ equations by suitable operations with the second equation, in its modified form with the coefficient of x_1 zero. The process is continued until the last equation is reached when the matrix has been reduced to upper triangular form. The right-hand sides of the equations are successively modified as the process continues and the unknowns finally evaluated by a 'back-substitution'. At the elimination of the ith unknown, the diagonal element in the ith equation is known as the 'pivot', and the method is sometimes called 'pivotal condensation'. In a straightforward application of the method the unknowns are eliminated in the order $1, 2, \dots, n$. Round-off errors can be reduced if, at each stage, the unknown to be eliminated is selected as the one with the numerically greatest diagonal element. This leads to more complicated programming with banded matrices and in the majority of structural problems a pivotal search is unnecessary, the straightforward method giving satisfactory results.

A flow diagram for the elimination process is shown in Fig. 7.3. The unknown being eliminated is indicated by the index i, the equation being operated on by j, and the element in the equation by k. It will be seen that the process requires a three-deep nest of cycles. The back-substitution may be dealt with in a similar way.

The application of the process to a symmetric band matrix requires a somewhat more sophisticated technique if full advantage is to be taken of the banded nature of the coefficient matrix. Consider the band matrix shown in Fig. 7.4. Only the upper part of the band, shown shaded, is stored in the computer. The coefficients in the lower half-band are implied from the relationship $a_{ij} = a_{ji}$. The elimination process, at each stage, need extend only to the lower edge of the band since the coefficients of the particular variable being eliminated are already zero in the subsequent equations.

A complete FORTRAN program for the solution of band matrices using this method is given in section 7.5.

Tri-diagonal recursion. In certain circumstances, the band matrix of coefficients contains non-zero elements on the leading diagonal and the adjacent diagonals only. Such a matrix was encountered in section 4.6, when considering the continuous beam, and is called a *tri-diagonal* matrix. The general

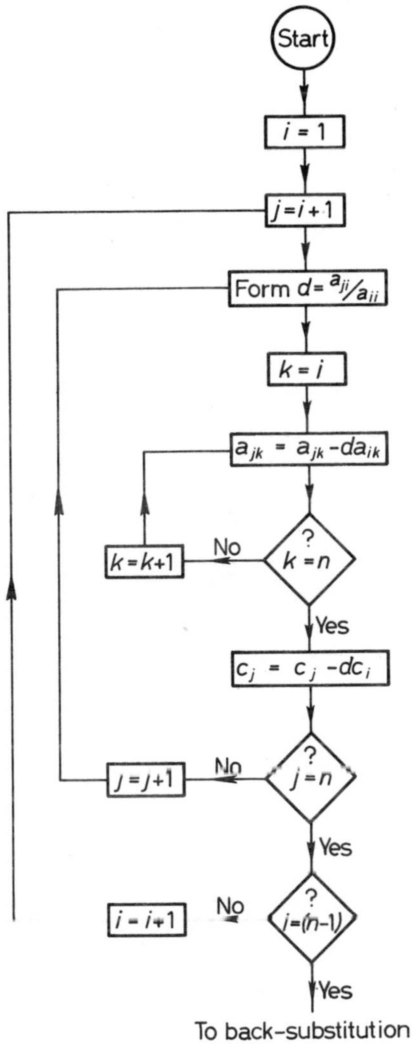

Figure 7.3. Flow diagram for Gauss elimination

form of the tri-diagonal matrix, in symmetric form, is

$$\begin{bmatrix} a_1 & b_2 & & & \\ b_2 & a_2 & b_3 & & \\ & b_3 & a_3 & \ddots & \\ & & \ddots & \ddots & \ddots \\ & & & \ddots & \ddots & b_n \\ & & & & b_n & a_n \end{bmatrix} \begin{bmatrix} x_1 \\ x_2 \\ x_3 \\ \vdots \\ x_{n-1} \\ x_n \end{bmatrix} = \begin{bmatrix} c_1 \\ c_2 \\ c_3 \\ \vdots \\ c_{n-1} \\ c_n \end{bmatrix} \tag{7.4}$$

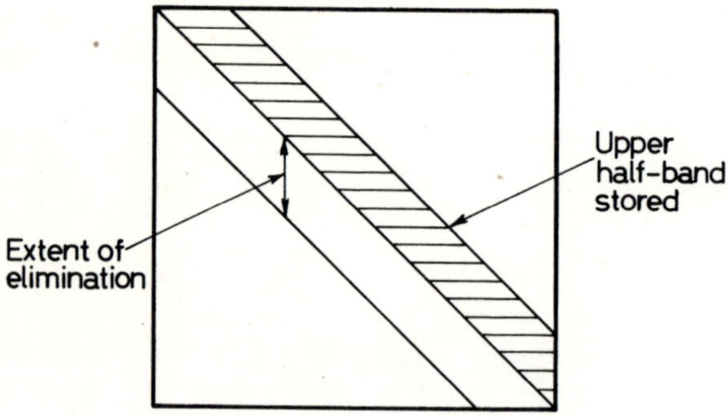

Figure 7.4. Storage of band matrix in elimination method

A simple algorithm for the solution of eqns. (7.4) may be established by successive elimination of the unknowns x_i as follows:

The first two equations are

$$\left.\begin{aligned} a_1x_1 + b_2x_2 &= c_1 \\ b_2x_1 + a_2x_2 + b_3x_3 &= c_2 \end{aligned}\right\} \tag{7.5}$$

Eliminating x_1 from eqns. (7.5)

$$x_2(a_2 - b_2^2/a_1) + b_3x_3 = c_2 - (b_2c_1/a_1) \tag{7.6}$$

Defining $\bar{a}_i = a_i - b_i^2/\bar{a}_{i-1}$

and $\bar{c}_i = c_i - b_i\bar{c}_{i-1}/\bar{a}_{i-1}$ $\Bigg\}$ $i = 2, 3, \ldots, n$ (7.7)

where $\bar{a}_1 = a_1$

and $\bar{c}_1 = c_1$

then eq. (7.6) becomes

$$\bar{a}_2x_2 + b_3x_3 = \bar{c}_2 \tag{7.8}$$

Equation (7.8) may now be used to eliminate x_2 from the third of eqns. (7.4), and the process is continued through the remaining equations until the last equation appears as

$$\bar{a}_n x_n = \bar{c}_n \tag{7.9}$$

from which

$$x_n = \bar{c}_n/\bar{a}_n \tag{7.10}$$

The back-substitution then proceeds as follows:

$$x_{n-1} = \bar{c}_{n-1}/\bar{a}_{n-1} - b_n x_n/\bar{a}_{n-1} \tag{7.11}$$

or, in general,

$$x_i = \bar{c}_i/\bar{a}_i - b_{i+1}x_{i+1}/\bar{a}_i \qquad i = (n-1), (n-2), \ldots, 1 \tag{7.12}$$

The algorithm may be summarized as follows:

$$\left.\begin{aligned}
\bar{a}_1 &= a_1 \\
\bar{c}_1 &= c_1 \\
\bar{a}_i &= a_i - b_i^2/\bar{a}_{i-1} \\
\bar{c}_i &= c_i - b_i\bar{c}_{i-1}/\bar{a}_{i-1} \quad & i = 2, 3, \ldots, n \\
x_n &= \bar{c}_n/\bar{a}_n \\
x_i &= \bar{c}_i/\bar{a}_i - b_{i+1}x_{i+1}/\bar{a}_i \quad & i = (n-1), (n-2), \ldots, 1
\end{aligned}\right\} \tag{7.13}$$

The method may be used in the case where the elements of the matrices in (7.4) are themselves matrices, the coefficient matrix being termed a quasi-tri-diagonal matrix. The form of the equations is

$$\begin{bmatrix}
\mathbf{A}_1 & \mathbf{B}_2 & & & \\
\mathbf{B}_2^T & \mathbf{A}_2 & \mathbf{B}_3 & & \\
 & \mathbf{B}_3^T & \mathbf{A}_3 & \ddots & \\
 & \ddots & \ddots & \ddots & \\
 & & \ddots & \ddots & \mathbf{B}_n \\
 & & & \mathbf{B}_n^T & \mathbf{A}_n
\end{bmatrix}
\begin{bmatrix} \mathbf{X}_1 \\ \mathbf{X}_2 \\ \mathbf{X}_3 \\ \vdots \\ \mathbf{X}_{n-1} \\ \mathbf{X}_n \end{bmatrix}
=
\begin{bmatrix} \mathbf{C}_1 \\ \mathbf{C}_2 \\ \mathbf{C}_3 \\ \vdots \\ \mathbf{C}_{n-1} \\ \mathbf{C}_n \end{bmatrix} \tag{7.14}$$

Applying a similar process of elimination, the resulting algorithm in this case is

$$\left.\begin{aligned}
\overline{\mathbf{A}}_1 &= \mathbf{A}_1 \\
\overline{\mathbf{C}}_1 &= \mathbf{C}_1 \\
\overline{\mathbf{A}}_i &= \mathbf{A}_i - \mathbf{B}_i^T\overline{\mathbf{A}}_{i-1}^{-1}\mathbf{B}_i \\
\overline{\mathbf{C}}_i &= \mathbf{C}_i - \mathbf{B}_i^T\overline{\mathbf{A}}_{i-1}^{-1}\overline{\mathbf{C}}_{i-1} \quad & i = 2, 3, \ldots, n \\
\mathbf{X}_n &= \overline{\mathbf{A}}_n^{-1}\overline{\mathbf{C}}_n \\
\mathbf{X}_i &= \overline{\mathbf{A}}_i^{-1}[\overline{\mathbf{C}}_i - \mathbf{B}_{i+1}\mathbf{X}_{i+1}] \quad & i = (n-1), (n-2), \ldots, 1
\end{aligned}\right\} \tag{7.15}$$

This algorithm is easily programmed for a digital computer. The **A** matrices may be successively overwritten by the $\bar{\mathbf{A}}$ matrices, and the **C** matrices may be successively overwritten by the $\bar{\mathbf{C}}$ and **X** matrices, thus the demands on computer storage are not excessive.

The Choleski method. In this method, the coefficient matrix is factorized in such a way that one of the factor matrices is upper triangular and the other lower triangular. Consider the system of equations

$$\mathbf{AX} = \mathbf{C} \tag{7.16}$$

and considering only the case where **A** is a symmetric matrix, we may put

$$\mathbf{A} = \mathbf{U}^T\mathbf{U} \tag{7.17}$$

in which **U** is an upper triangular matrix.
Substituting eq. (7.17) in eq. (7.16),

$$\mathbf{U}^T\mathbf{UX} = \mathbf{C} \tag{7.18}$$

Putting

$$\mathbf{UX} = \mathbf{Y} \tag{7.19}$$

then

$$\mathbf{U}^T\mathbf{Y} = \mathbf{C} \tag{7.20}$$

Having found the upper triangular factor matrix **U**, the process consists of solving eq. (7.20) for **Y** and then solving eq. (7.19) for **X**.

If the coefficient matrix **A** is a band matrix, of semi-band width B then it is not difficult to show that the factor matrices **U** and $\mathbf{U}^T$ are also band matrices with a band width B. Thus, eq. (7.17) has the form shown in Fig. 7.5.

The elements of the matrix **U**, and hence $\mathbf{U}^T$, may be computed, using the rule for matrix multiplication, as follows:

$$\left.\begin{aligned}
u_{11} &= \sqrt{a_{11}} \\
u_{1i} &= a_{1i}/u_{11} \\
u_{ii} &= \sqrt{\left(a_{ii} - \sum_{m=1}^{(i-1)} u_{mi}{}^2\right)}; \quad i > 1 \\
u_{ij} &= \left(a_{ij} - \sum_{m=1}^{(i-1)} u_{mi}u_{mj}\right)\Bigg/ u_{ii}; \quad j > i \\
u_{ij} &= 0; \quad i > j
\end{aligned}\right\} \tag{7.21}$$

Only the upper half-band of the **A** matrix is stored as a one-dimensional array. Thus, the element ij of the two-dimensional array becomes element

$$[(i - 1)B + j - i + 1]$$

in the one-dimensional array.

The evaluation of **Y** from eq. (7.20) proceeds as follows:

$$\left.\begin{aligned} y_1 &= c_1/u_{11} \\ y_i &= \left(c_i - \sum_{m=1}^{(i-1)} u_{mi} y_m\right) \Big/ u_{ii}; \quad i > 1 \end{aligned}\right\} \tag{7.22}$$

The solutions **X** are then found by solving eq. (7.19), commencing the process with the last equation.

$$\left.\begin{aligned} x_n &= y_n/u_{nn} \\ x_i &= \left(y_i - \sum_{m=i+1}^{n} u_{im} x_m\right) \Big/ u_{ii}; \quad i < n \end{aligned}\right\} \tag{7.23}$$

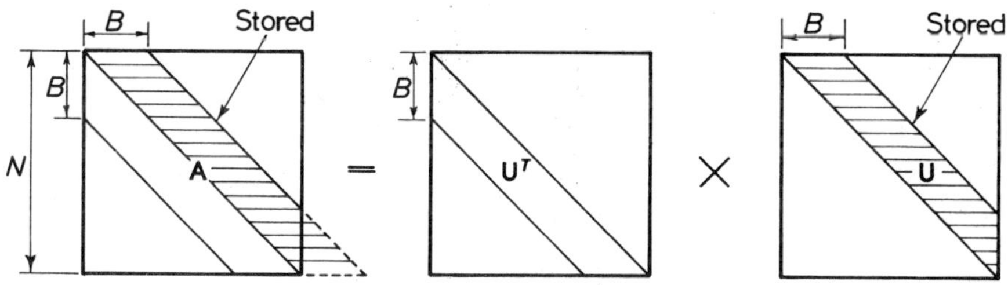

Figure 7.5. Storage arrangement in Choleski method

A complete program, coded in ALGOL, for this method is given in section 7.6, along with a specification of the data. The elements of the matrix **A** are overwritten by the elements of **U** as they are calculated, and the elements of **C** are successively overwritten by the elements of the intermediate vector **Y** and the solution vector **X**. Thus, the storage required is only $N(B + 1)$, plus a small amount of space for the constants.

7.5. FORTRAN Program for Band Matrix Solution by Gauss Elimination

```
GO #XFAP 20
SHORTLIST (TP)
SEND TO(TP)
PROGRAM(DASW)
INPUT 1=TRO
OUTPUT 2=TPO
END
MASTER SOLUTION
DIMENSION F(1,132),C(1,2592),A(1,144)
READ(1,108)NB,NO
```

```
108   FORMAT(2I0)
      NOB=NB*NO
      NONB=NB+NO
```

Read coefficient matrix

```
      DO 50 I=1,NOB
      READ(1,109)C(1,I)
109   FORMAT(F0.0)
50    CONTINUE
```

Read applied load vector

```
      DO 51 I=1,NONB
      READ(1,110)F(1,I)
110   FORMAT(F0.0)
51    CONTINUE
```

Eliminate variables from equations 1 to $(N-B+1)$

```
      LA=NB-1
      LB=NO-LA
      DO 72 I=1,LB
      J=NB*I-LA
      DO 73 K=1,LA
      A(1,K)=C(1,J+K)
73    CONTINUE
      DO 74 K=1,LA
      C(1,J+K)=C(1,J+K)/C(1,J)
74    CONTINUE
      F(1,I)=F(1,I)/C(1,J)
      LC=LA
      DO 75 L=1,LA
      MI=J+L*NB-1
      Z=F(1,I)
      X=A(1,L)
      F(1,I+L)=F(1,L+I)-Z*X
      DO 76 K=1,LC
      Z=C(1,J+K+LA-LC)
      X=A(1,L)
      C(1,MI+K)=C(1,MI+K)-Z*X
76    CONTINUE
      LC=LC-1
75    CONTINUE
72    CONTINUE
```

Eliminate variables from equations $(N-B+2)$ to N

```
      LB=LB+1
      LD=LA-1
      DO 77 I=LB,NO
      J=NB*I-LA
      IF(1-LD)300,300,301
300   DO 78 K=1, LD
      A(1,K)=C(1,J+K)
78    CONTINUE
301   DO 79 K=1,LA
      C(1,J+K)=C(1,J+K)/C(1,J)
79    CONTINUE
      F(1,I)=F(1,I)/C(1,J)
      LC=LA-1
      IF(I-NO)302,303,302
302   DO 305 L=1,LD
      MI=L*NB+J-1
```

```
      Z=F(1,I)
      X=A(1,L)
      F(1,L+I)=F(1,L+I)-Z*X
      DO 81 K=1,LC
      Z=C(1,J+K+LA-LC-1)
      X=A(1,L)
      C(1,MI+K)=C(1,MI+K)-Z*X
81    CONTINUE
      LC=LC-1
      IF(1-LC) 305,305,304
305   CONTINUE
304   LD=LD-1
77    CONTINUE
303   DO 83 IK=1,NO
      I=NO-IK+1
      J=NB*I-NB+2
      DO 82 K=1,LA
      Z=F(1, I+K)
      X=C(1,J+K-1)
      F(1,I)=F(1,I)-Z*X
82    CONTINUE
83    CONTINUE
      WRITE(2,206)(F(1,K),K=1,NO)
206   FORMAT(E14.7)
      STOP
      END
      FINISH
      ####
```

Eliminate variables from equations $(N-B+2)$ to N

Back-substitution

Print solutions

Arrangement of data for Gauss elimination program.

The program calls for data in the following order:

B, the half-band width, Fig. 7.6;

N, the total number of equations;

A, the matrix of coefficients in semi-band form, punched by rows with each element on a new line. A dummy triangle of zeros must be included at the bottom of the band, as shown shaded in Fig. 7.6. Thus, the total number of elements in the array **A** is BN;

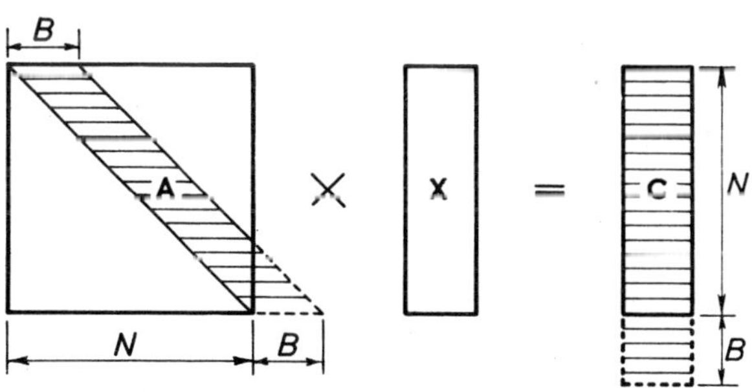

Figure 7.6. Storage arrangement in elimination method

C, the right-hand side column matrix consisting of N elements, each on a new line, and with B zeros added at the end.

The data is prepared in normal 'fixed point' form and the output, simply the solution vector **X**, is in 'floating point' form.

7.6. ALGOL Program for Band Matrix Solution by the Choleski Method

```
GO #XASP21
'PROGRAM' (GJH5)
'INPUT' 0=TR0
'OUTPUT' 0=TP0

'BEGIN'

'PROCEDURE' FACTOR (K,B,N);
'VALUE' B,N;
'ARRAY' K;
'INTEGER' B,N;
'BEGIN'
'INTEGER' I,J,L,Q;
'REAL' E;
'BOOLEAN' IB,JB;
K[1]:=SQRT(K[1]);
'FOR' J:=2 'STEP' 1 'UNTIL' B 'DO'
K[J]:=K[J]/K[1];
'FOR' I:=2 'STEP' 1 'UNTIL' N 'DO'
'BEGIN'
IB:=I 'LE' B;
'FOR' J:=I 'STEP' 1 'UNTIL' I+B-1 'DO'
'BEGIN'
JB:=J 'LE' B;
'IF' J>N 'THEN' 'GOTO' L1;
'IF' I=J 'THEN'
'BEGIN'
Q:='IF' IB 'THEN' 1 'ELSE' I-B+1;
E:=0;
'FOR' L:=Q 'STEP' 1 'UNTIL' I-1 'DO'
E:=E+K[(L-1)*B+I+1-L]↑ 2;
K[(I-1)*B+1]:=SQRT(K[(I-1)*B+1]-E)
'END'
'ELSE'
'BEGIN'
Q:='IF' JB 'THEN' 1 'ELSE' J-B+1;
E:=0;
'FOR' L:=Q 'STEP' 1 'UNTIL' I-1 'DO'
E:=E+K[(L-1)*B+I+1-L]*K[(L-1)*B+J+1-L];
K[(I-1)*B+J+1-I]:=(K[(I-1)*B+J+1-I]-E)/K[(I-1)*B+1]
'END'
'END'
L1: 'END'
'END';
```

Computation of factor matrix **U**

First row of **U**

$\mathbf{U}_{ii}$ for $i=2(1)N$

$\mathbf{U}_{ij}$ for $i=2(1)N$, $j=i(1)i+B+1$

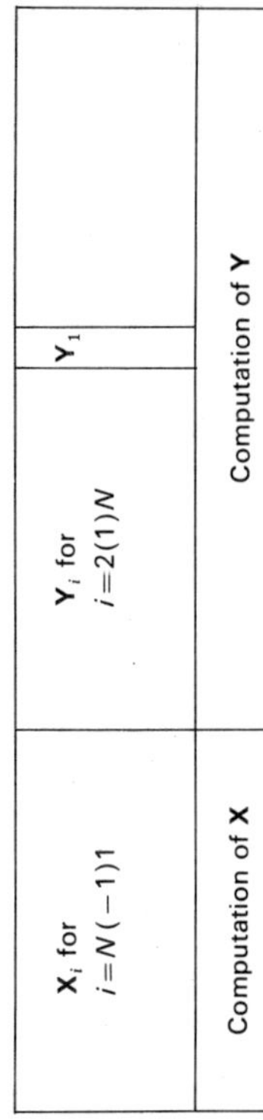


```
'PROCEDURE' BACKSUB(K,R,B,N);
'VALUE' K,B,N;
'ARRAY' K,R;
'INTEGER' B,N;
'BEGIN'
'INTEGER' I,L,Q;
'REAL' E;
'BOOLEAN' IB;
R[1]:=R[1]/K[1];
'FOR' I:=2 'STEP' 1 'UNTIL' N 'DO'
'BEGIN'
IB:=I 'LE' B;
Q:='IF' IB 'THEN' 1 'ELSE' I-B+1;
E:=0;
'FOR' L:=Q 'STEP' 1 'UNTIL' I-1 'DO'
E:=E+K[(L-1)*B+I+1-L]*R[L];
R[I]:=(R[I]-E)/K[(I-1)*B+1]
'END';
R[N]:=R[N]/K[(N-1)*B+1];
'FOR' I:=N-1 'STEP'-1 'UNTIL' 1 'DO'
'BEGIN'
Q:='IF' I+B-1 'GE' N 'THEN' N 'ELSE' I+B-1;
E:=0;
'FOR' L:=I+1 'STEP' 1 'UNTIL' Q 'DO'
E:=E+K[(I-1)*B+L+1-I]*R[L];
R[I]:=(R[I]-E)/K[(I-1)*B+1]
'END'
'END';

'INTEGER' NSYSTEM,COUNT;
NSYSTEM:=READ;
'FOR' COUNT:=1 'STEP' 1 'UNTIL' NSYSTEM 'DO'
'BEGIN'

'INTEGER' HBW, NEQN, NRHS;
NEQN:=READ;
HBW:=READ;
NRHS:=READ;
'BEGIN'

'ARRAY' COEFF[1:HBW*NEQN],RHS[1:NEQN];
'INTEGER' I,J;
'FOR' I:=1 'STEP' 1 'UNTIL' HBW*NEQN 'DO'
COEFF[I]:=READ;
FACTOR(COEFF, HBW, NEQN);
'FOR' J:=1 'STEP' 1 'UNTIL' NRHS 'DO'
'BEGIN'
'FOR' I:=1 'STEP' 1 'UNTIL' NEQN 'DO'
RHS[I]:=READ;
BACKSUB(COEFF,RHS,HBW,NEQN);
NEWLINE(10);
```

```
'FOR' I:=1 'STEP' 1 'UNTIL' NEQN 'DO'
'BEGIN'
PRINT(I,3,0);
PRINT(RHS[I],0,10);
NEWLINE(1);
'END'
'END'
'END'
'END'

'END'

####
```

Print solution

Arrangement of data—Choleski method.

The program calls for data in the following order:

NSYSTEM = number of sets to be solved;

NEQN = number of equations;

HBW = half-band width;

NRHS = number of right-hand sides.

Coefficient matrix in semi-band form, punched by rows and having a dummy triangle of zeros included at the bottom of the band to preserve the band width (HBW × NEQN numbers).

The right-hand side column matrix (repeated NRHS times).

The bracketed items of data above are repeated NSYSTEM times.

The output format is in 'floating point' form, the NEQN numbers being printed for each right-hand side for each set of NSYSTEM.

8. Computer Applications

In this chapter we shall consider some examples of computer applications in structural analysis. The first two examples deal with non-prismatic members in structures, and computer programs will be presented for the calculation of end stiffnesses, of an isolated non-prismatic member, and for the complete analysis of a structure using the flexibility method. A further application of the flexibility method will be briefly outlined in section 8.3, in the analysis of a cable-braced girder.

As an illustration of a structure suitable for analysis by the finite element method, section 8.4 will be devoted to the analysis of a perforated shear wall of the type found in tall buildings. A rectangular plane stress element, of the type derived in section 6.4, is suitable for this purpose.

8.1. A Program to Calculate End Stiffnesses of a Non-prismatic Member

A suitable method for carrying out this calculation on a computer is the one discussed in section 5.1, in which the end flexibilities are calculated, using Simpson's rule to effect the numerical integration, and the stiffnesses obtained from these. The relevant expressions for the end flexibilities are given in eq. (5.1) and for the stiffnesses in eqns. (5.2) to (5.5).

In order to make this first example as simple as possible, the length of the member is divided into a fixed number of steps for the application of Simpson's rule. The method could, without difficulty, be programmed for a variable number of steps and this would be a useful exercise for the reader. In the majority of cases, ten divisions will give sufficient accuracy and this number has been used in the program.

The ordinates of the m_1 and m_2 diagrams, Fig. 8.1, are simply 0, 0·1, 0·2, . . ., 1 and the step length h is constant at $l/10$, where l is the length of the member.

The values of $m_1{}^2/EI$, $m_2{}^2/EI$, and m_1m_2/EI are calculated at each of the stations numbered 1 to 11. Simpson's rule is then used to obtain the integrals and a flow diagram for this is shown in Fig. 8.2. The ordinates of the function being integrated are designated F and the result of the integration is T. A more compact programming of Simpson's rule is available in the ALGOL program in section 8.2, in which the multiplications are performed by a single, compound, instruction.

The data consists of the *EI* values, either relative or absolute values may

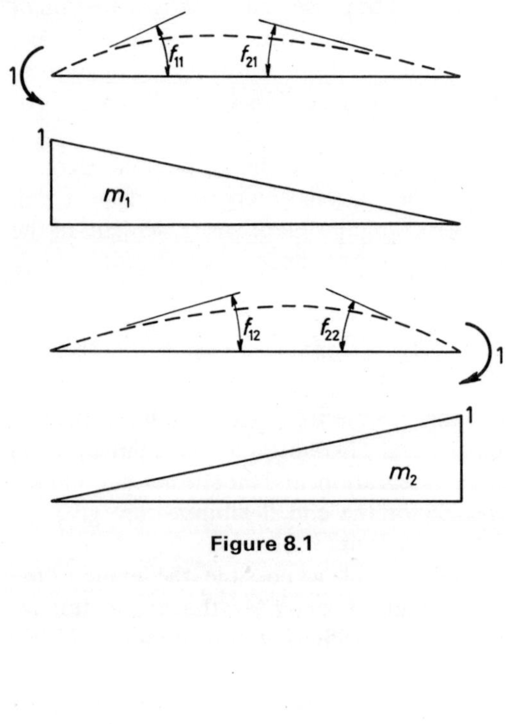

Figure 8.1

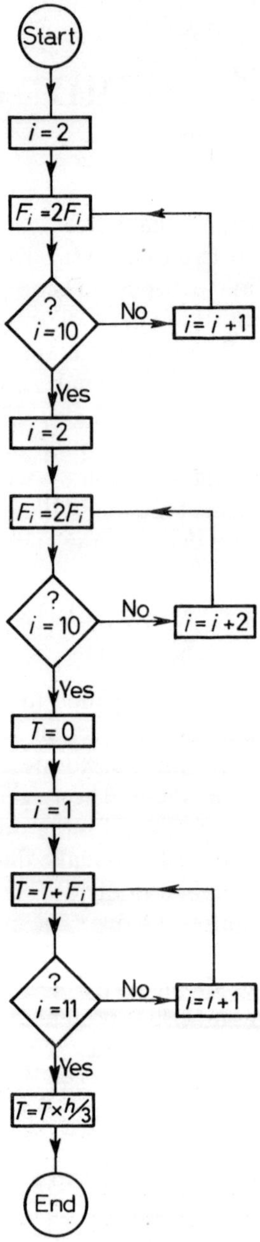

Figure 8.2. Flow diagram for Simpson's Rule

be used, at the eleven stations, and the length of the member. The output consists of the stiffnesses k_{11} and k_{22} and the carry-over factors C_{12} and C_{21}.

Fortran program for non-prismatic member stiffnesses.

```
      GO #XFAP 20
      SHORTLIST(TP)
      SEND TO(TP)
      PROGRAM(WJMS)
      INPUT 1=TRO
      OUTPUT 2=TPO
      END

      MASTER WJMS
      WRITE(2,100)
100   FORMAT(/33H NON PRISMATIC MEMBER STIFFNESSES/)
      REAL L,M1,M2,M12,K1,K2
      DIMENSION EI(11), M1(11), M2(11), M12(11)
      DO1 I=1,11
      M1(I)=1.0-0.1*(I-1)                           Set m1
1     M2(I)=0.1*(I-1)                               Set m2
      READ(1,101)(EI(I),I=1,11)                     Read EI values
101   FORMAT(11F0·0)
      READ(1,102)L                                  Read l
102   FORMAT(1F0.0)
      DO2 I=1,11
      M12(I)=M1(I)*M2(I)/EI(I)                      m1m2/EI
      M1(I)=M1(I)**2/EI(I)                          m1^2/EI

2     M2(I)=M2(I)**2(EI(I)                          m2^2/FI
      DO3 I=2,10                                    Application of
      M1(I)=2*M1(I)                                 Simpson's rule
      M2(I)=2*M2(I)
3     M12(I)=2*M12(I)
      DO4 I=2,10,2
      M1(I)=2*M1(I)
      M2(I)=2*M2(I)
4     M12(I)=2*M12(I)
      F1=0
      F2=0
      F12=0
      DO5 I=1,11
      F1=F1+M1(I)
      F2=F2+M2(I)
5     F12=F12+M12(I)
      F1=F1*0.1*L/3
      F2=F2*0.1*L/3
      F12=F12*0.1*L/3
      K1=F2/(F1*F2-F12**2)                          k1
      K2=F1/(F1*F2-F12**2)                          k2
      C1=F12/F2                                     c1
      C2=F12/F1                                     c2
      WRITE(2,103)K1,K2,C1,C2                       Print results
```

```
103 FORMAT(4E15.5)
    STOP
    END
    FINISH
    ####
```

8.2. Influence Line Calculations for Structures with Non-prismatic Members

The analysis of structures with non-prismatic members, by the methods discussed in previous chapters, presents no difficulty, but does tend to require a substantial amount of computation. This is particularly evident if influence line ordinates are required. This type of structure is therefore one for which the digital computer has much to offer.

The method which will now be outlined is based on the flexibility concept and a computer program in ALGOL will be given at the end of this section. The dimensions of the matrices involved in the calculation can all be defined by three program parameters, N the number of redundants, W the number of loading systems (i.e., the number of unit load positions if influence line ordinates are required), and V the total number of integration stations used in the Simpson's rule integrations of eqns. (5.1). The compatibility conditions are assembled in the form, eq. (2.12),

$$\mathbf{FX} + \mathbf{U} = \mathbf{0}$$

in which the matrix dimensions are $\mathbf{F}\,(N \times N)$, $\mathbf{X}\,(N \times W)$, and $\mathbf{U}\,(N \times W)$.

We shall now consider the computation of the matrices **F** and **U** from given structural data.

Computation of F. The elements of the flexibility matrix **F** are given by

$$f_{rs} = \int \frac{m_r m_s \, \mathrm{d}s}{EI} \tag{8.1}$$

Bending moment distributions m are obtained for the primary structure for unit values of each of the N redundancies. A matrix **M** is then defined as follows:

$$\underset{(V \times N)}{\mathbf{M}} = \begin{bmatrix} (m_1)_1 & (m_2)_1 & \dots & (m_N)_1 \\ (m_1)_2 & (m_2)_2 & \dots & (m_N)_2 \\ \vdots & \vdots & & \vdots \\ (m_1)_V & (m_2)_V & \dots & (m_N)_V \end{bmatrix} \tag{8.2}$$

A typical element $(m_i)_j$ of the matrix **M** is the value of bending moment in the structure at the jth integration station, due to unit value of the ith redundant. The integrations required by equations of type (8.1) can be performed by matrix multiplications, if we define a diagonal matrix **D** in which the diagonal elements are

$$\frac{h}{3EI} \times \text{(Simpson's rule coefficient)}$$

at each of the V integration stations where h is the appropriate step length. It is unnecessary to store **D** as a diagonal matrix in the computer and in the program the diagonal elements are simply stored as a row matrix. However, for the purposes of outlining the method, it is convenient to regard **D** as a diagonal matrix.

Pre-multiplication of **M** by **D** gives

$$\mathbf{DM} = \begin{bmatrix} D_1(m_1)_1 & D_1(m_2)_1 & \cdots & D_1(m_N)_1 \\ D_2(m_1)_2 & D_2(m_2)_2 & \cdots & D_2(m_N)_2 \\ \vdots & \vdots & & \vdots \\ D_V(m_1)_V & D_V(m_2)_V & \cdots & D_V(m_N)_V \end{bmatrix} \tag{8.3}$$

DM is then transposed and post-multiplied by **M**, thus

$$(\mathbf{DM})^T\mathbf{M} = \begin{bmatrix} \sum_{p=1}^{V} (Dm_1m_1)_p & \sum_{p=1}^{V} (Dm_1m_2)_p & \cdots & \sum_{p=1}^{V} (Dm_1m_N)_p \\ \sum_{p=1}^{V} (Dm_2m_1)_p & \sum_{p=1}^{V} (Dm_2m_2)_p & \cdots & \sum_{p=1}^{V} (Dm_2m_N)_p \\ \vdots & \vdots & & \vdots \\ \sum_{p=1}^{V} (Dm_Nm_1)_p & \sum_{p=1}^{V} (Dm_Nm_2)_p & \cdots & \sum_{p=1}^{V} (Dm_Nm_N)_p \end{bmatrix} \tag{8.4}$$

Thus, comparing eqns. (8.1) and (8.4),

$$\mathbf{F} = (\mathbf{DM})^T\mathbf{M} \tag{8.5}$$

Computation of U. The elements of **U** are given by

$$u_{ij} = \int \frac{m_i(m_0)_j \, ds}{EI} \tag{8.6}$$

in which u_{ij} is the primary structure discontinuity at the ith redundant due to the jth loading system.

A matrix $\mathbf{M}_0$ is introduced and defined as follows:

$$\underset{(V \times W)}{\mathbf{M}_0} = \begin{bmatrix} (m_0)_{1,1} & (m_0)_{1,2} & \cdots & (m_0)_{1,W} \\ (m_0)_{2,1} & (m_0)_{2,2} & \cdots & (m_0)_{2,W} \\ \vdots & \vdots & & \vdots \\ (m_0)_{V,1} & (m_0)_{V,2} & \cdots & (m_0)_{V,W} \end{bmatrix} \tag{8.7}$$

A typical element $(m_0)_{i,j}$ of the matrix $\mathbf{M}_0$ is the value of bending moment in the primary structure at the *i*th integration station, due to the *j*th applied load system.

The integrations required by eq. (8.6) are then obtained by matrix multiplications as follows:

$$(\mathbf{DM})^T\mathbf{M}_0 = \begin{bmatrix} \sum_{p=1}^{V}[Dm_1(m_0)_1]_p & \sum_{p=1}^{V}[Dm_1(m_0)_2]_p & \cdots & \sum_{p=1}^{V}[Dm_1(m_0)_W]_p \\ \sum_{p=1}^{V}[Dm_2(m_0)_1]_p & \sum_{p=1}^{V}[Dm_2(m_0)_2]_p & \cdots & \sum_{p=1}^{V}[Dm_2(m_0)_W]_p \\ \vdots & \vdots & & \vdots \\ \sum_{p=1}^{V}[Dm_N(m_0)_1]_p & \sum_{p=1}^{V}[Dm_N(m_0)_2]_p & \cdots & \sum_{p=1}^{V}[Dm_N(m_0)_W]_p \end{bmatrix} \tag{8.8}$$

Thus, comparing eqns. (8.6) and (8.8),

$$\mathbf{U} = (\mathbf{DM})^T\mathbf{M}_0 \tag{8.9}$$

The solutions are obtained as

$$\mathbf{X} = -\mathbf{F}^{-1}\mathbf{U} = -[(\mathbf{DM})^T\mathbf{M}]^{-1}[(\mathbf{DM})^T\mathbf{M}_0] \tag{8.10}$$

For influence line computations the number of unit load positions will generally be much greater than the number of redundancies, and thus it is economical to form the inverse flexibility matrix. This is the course adopted in the program.

Preparation of data. The assembly of all the matrices required is carried out automatically by the program, reducing the amount of data required to a minimum. The first items of data are: M the number of members, N the number of redundants, V the number of integration stations, and W the number of loading systems. There follows a list of the EI values at the integration stations, the step length for each member, and the number of stations in each member, and the latter must of course be odd numbers to satisfy the Simpson's rule requirement. The remainder of the data consists of sufficient information to define the m and m_0 bending moment distributions. Three types of bending moment diagram can be accepted for each member. Each diagram is given a type number and this is followed by the data required to define the particular diagram. The three types of diagram with the type numbers and data required are shown in Table 8.1. An example set of data is assembled in Table 8.2, which represents the data relating to the structure shown in Fig. 8.3. It should be noted that the number of stations, V, is the total for all members, i.e., end stations are included for each member meeting at a joint.

Table 8.1

Diagram Type	Type number	Additional data
	0	Nil
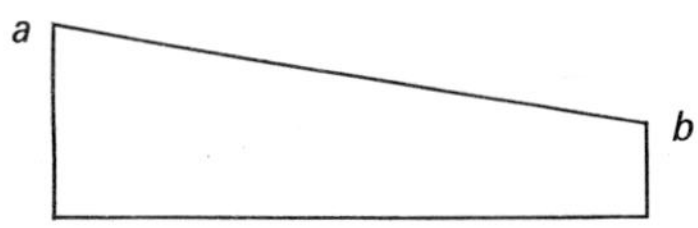	1	*a*, *b*
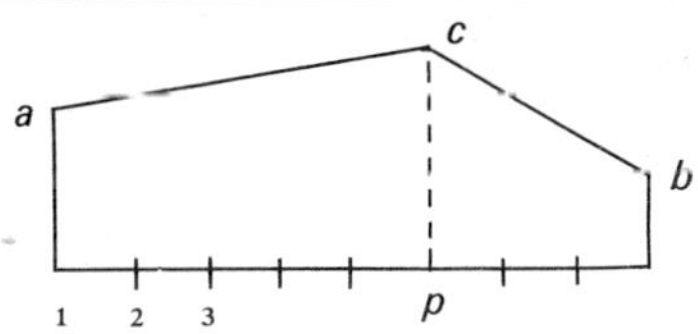	2	*a*, *b*, *c*, *p*

A discontinuity in the *EI* variation, or in one of the bending moment diagrams, will introduce an error if it occurs at one of the even-numbered points. The difficulty can be avoided if the intervals are chosen so that the discontinuity falls at an odd-numbered station. Thus, in the example in Fig. 8.3, the unit load positions are chosen to correspond with alternate stations and thus automatically lie at odd-numbered points.

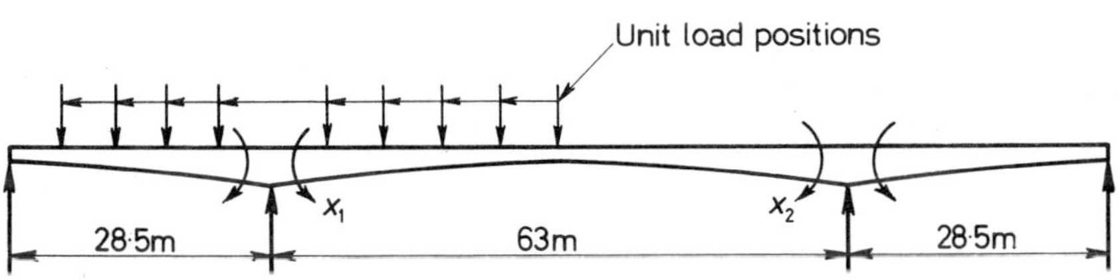

Figure 8.3

The program is designed specifically to deal with non-prismatic members, but it can, of course, be used in cases where some or all of the members are prismatic, although in the latter case it will not be the most economical method.

The program, coded in ALGOL, is now given in full. It will be noticed that the program consists in the main of four 'procedures': AMAT (used for matrix assembly), MATMULT (used for matrix multiplications), MATINV (an external procedure for matrix inversion), and MATPRINT (for printing the results). The procedures are entered as required during the execution of the program.

Table 8.2. Data for Example of Fig. 8.3

3							M		
2							N		
43							V		
9							W		
410	430	470	510	610	750	920	$EI \times 10^{-9}$ (kNmm^2)		
1130	1400	1790	2240				$EI \times 10^{-9}$ (kNmm^2)		
2240	1850	1520	1280	1080	940	850	$EI \times 10^{-9}$ (kNmm^2)		
780	720	670	667	670	720	780	$EI \times 10^{-9}$ (kNmm^2)		
850	940	1080	1280	1520	1850	2240	$EI \times 10^{-9}$ (kNmm^2)		
2240	1790	1400	1130	920	750	610	$EI \times 10^{-9}$ (kNmm^2)		
510	470	430	410				$EI \times 10^{-9}$ (kNmm^2)		
2·85	3·15	2·85					h		
11	21	11					s (number of stations/member)		
1	0	1					Span 1	x_1	m
1	1	0					Span 2	x_1	m
0							Span 3	x_1	m
0							Span 1	x_2	m
1	0	1					Span 2	x_2	m
1	1	0					Span 3	x_2	m
2	0	0	−4·56	3	0	0	m_0		
2	0	0	−6·84	5	0	0	m_0		
2	0	0	−6·84	7	0	0	m_0		
2	0	0	−4·56	9	0	0	m_0		
0	2	0	0	−5·67	3	0	m_0		
0	2	0	0	−10·08	5	0	m_0		
0	2	0	0	−13·23	7	0	m_0		
0	2	0	0	−15·12	9	0	m_0		
0	2	0	0	−15·75	11	0	m_0		

ALGOL program for influence line calculations for structures with non-prismatic members

```
GO #XASP 21
'PROGRAM' (WJIL)
'INPUT' 0=TR0
'OUTPUT' 0=TP0
'BEGIN' 'INTEGER' M,N,V,W;
        M:=READ;
        N:=READ;
        V:=READ;
        W:=READ;
'BEGIN' 'INTEGER' P,Z,I,J,K,L,T;
        'ARRAY' D[1:V],BM[I:V,1:W],DMT[1:N,1:V],F[1:N,1:N],U,R[1:N,1:W],
            H[1:M];
        'INTEGER' 'ARRAY' S[1:M];

        'PROCEDURE' AMAT(C); 'VALUE' C; 'INTEGER' C;
'BEGIN' 'SWITCH' BMTYPE:=L1,L2,L3;
            'REAL' X,Y,Q;
        'FOR' I:=1 'STEP' 1 'UNTIL' V 'DO'
        'FOR' J:=1 'STEP' 1 'UNTIL' C 'DO' BM[I,J]:=0.0;
'FOR' I:=1 'STEP' 1 'UNTIL' C 'DO'
'BEGIN' T:=0;
        'FOR' J:=1 'STEP' 1 'UNTIL' M 'DO'
'BEGIN' Z:=READ;
        Z:=Z+1;
        'GOTO' BMTYPE[Z];
        L1:
        'GOTO' L4;
        L2: X:=READ;
            Y:=READ;
        'FOR' K:=1 'STEP' 1 'UNTIL' S[J] 'DO'
        BM[(T+K),I]:=(Y-X)*(K-1)/(S[J]-1)+X;
        'GOTO' L4;
        L3: X:=READ;
        Y: =READ;
        Q:=READ;
        P: =READ;
        'FOR' K:=1 'STEP' 1 'UNTIL' P 'DO'
        BM[(T+K),I]:=(Q-X)*(K-1)/(P-1)+X;
        'FOR' K:=(P+1) 'STEP' 1 'UNTIL' S[J] 'DO'
        BM[(T+K),I]:=(Y-Q)*(K-P)/(S[J]-P)+Q;
        L4: T:=T+S[J];
'END'   OF MEMBER
'END'   OF REDUNDANT(OR LOAD)
'END';
        'PROCEDURE' MATMULT(E,F,G,I,J,K); 'VALUE' I,J,K;
        'INTEGER' I,J,K; 'ARRAY' E,F,G;
'BEGIN' 'INTEGER' P,Q,R;
        'FOR' P:=1 'STEP' 1 'UNTIL' I 'DO'
        'FOR' Q:=1 'STEP' 1 'UNTIL' K 'DO'
'BEGIN' E[P,Q]:=0;
        'FOR' R:=1 'STEP' 1 'UNTIL' J 'DO'
        E[P,Q]:=E[P,Q]+F[P,R]*G[R,Q];
'END'
'END';
```

Procedure for matrix assembly

I = Subscript for redundant or load

T = Gives control on row

J = Subscript for member

Z = Diagram type number

Procedure for matrix multiplication

$\mathbf{E}(I \times K) = \mathbf{F}(I \times J) \times \mathbf{G}(J \times K)$

```
        'PROCEDURE' MATINV (A,N); 'VALUE' N;              | External procedure
        'ARRAY' A; 'INTEGER' N;                            | for matrix inversion
        'EXTERNAL';

        'PROCEDURE' MATPRINT (NAME,R,C);                   | Procedure for
        'VALUE' R,C,NAME;                                  | printing matrices
        'INTEGER' R,C;
        'ARRAY' NAME;
'BEGIN' 'INTEGER' I,J;
        'FOR' J:=1 'STEP' 1 'UNTIL' C 'DO'
'BEGIN' NEWLINE(4);
        WRITE TEXT('('COL.%NO.')');
        PRINT(J,3,0);
        NEWLINE(2);
        'FOR'I:=1 'STEP' 1 'UNTIL' R 'DO'
        OUTPUT(NAME[I,J]);
'END'
'END';

        'FOR' I:=1 'STEP' 1 'UNTIL' V 'DO' D[I]:=READ;     | Read EI
        'FOR' I:=1 'STEP' 1 'UNTIL' M 'DO' H[I]:=READ;     | Read h
        'FOR' I:=1 'STEP' 1 'UNTIL' M 'DO' S[I]:=READ;     | Read s
        L:=0;                                              | Assemble D
        'FOR' I:=1 'STEP' 1 'UNTIL' M 'DO'
        'FOR' J:=1 'STEP' 1 'UNTIL' S[I] 'DO'
'BEGIN' L:=L+1;
        D[L]:=H[I]/(3*D[L])*('IF' J=1 'OR' J=S[I] 'THEN'
           1 'ELSE' 'IF' (-1)↑J<0 'THEN' 2 'ELSE' 4);
'END';

        AMAT (N);                                          | Assemble M
        'FOR' I:=1 'STEP' 1 'UNTIL' V 'DO'                 | Form (DM)^T
        'FOR' J:=1 'STEP' 1 'UNTIL' N 'DO'
        DMT[J,I]:=D[I]*BM[I,J];

        MATMULT(F,DMT,BM,N,V,N);                           | (DM)^T M = F
        AMAT(W);                                           | Assemble M_0

        MATMULT(U,DMT,BM,N,V,W);                           | (DM)^T M_0 = U

        MATINV(F,N);                                       | F^-1
        MATPRINT(F,N,N);                                   | Print F^-1
        MATMULT(R,F,U,N,N,W);                              | -X = F^-1 U
        'FOR' I:=1 'STEP ' 1 'UNTIL' N 'DO'                | Print X
        'FOR' J:=1 'STEP' 1 'UNTIL' W 'DO' R[I,J]:=-R[I,J];
        MATPRINT(R,N,W);

'END'
'END'
####
```

8.3. The Cable-stayed Bridge—Flexibility Method

The cable-stayed girder behaves as a continuous beam supported elastically at the points of attachment of the cables. The flexibility method is suitable for analysis since the number of redundancies is usually not large and the flexural and direct flexibilities are easily evaluated and added to give the total flexibility.

A 'mixed' method of analysis, in which both forces and displacements constitute the unknowns, has been outlined by Stafford Smith.[33, 34]

Consider the cable-stayed girder shown in Fig. 8.4, in which the back-stay AF is attached to the girder at end A and the fore-stays FC and FD are attached to the girder at C and D. The tower BF is pinned at B and the girder is simply-supported at A, B, and E. The structure has three redundancies,

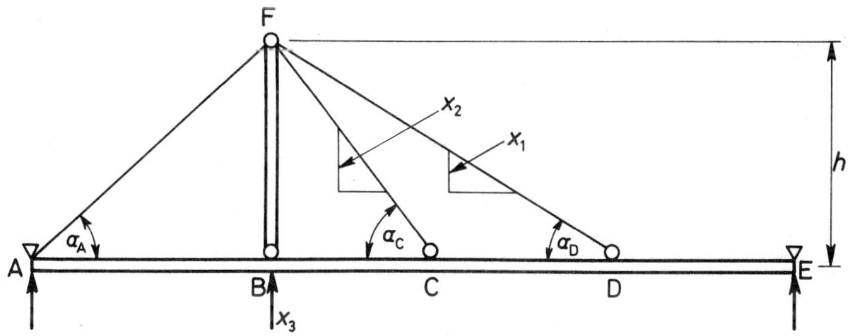

Figure 8.4. Cable stayed bridge

which are conveniently taken to be the vertical components of the tensions in the cables FC and FD and the vertical reaction at B. All the cables are pin-connected to the top of the tower F.

In applying the flexibility method, consideration is given to the deflections of the girder at points B, C, and D and to the deflections of the cable ends C and D separately. The compatibility conditions to be imposed are the final zero deflection of the girder at B and the equality of girder and cable deflections at C and D.

Consider the compatibility condition at D in the vertical direction:

$$\left.\begin{aligned} f_{11}x_1 &= \text{deflection of girder at D due to } x_1 \\ f_{12}x_2 &= \quad ,, \quad ,, \quad ,, \quad ,, \text{ D} \; ,, \; ,, \; x_2 \\ f_{13}x_3 &= \quad ,, \quad ,, \quad ,, \quad ,, \text{ D} \; ,, \; ,, \; x_3 \\ u_1 &= \quad ,, \quad ,, \quad ,, \quad ,, \text{ D} \; ,, \; ,, \text{ applied loads} \end{aligned}\right\} \quad (8.11)$$

For compatibility at D, the sum of the deflections, eq. (8.11), must equal the displacement of the cable end D in the actual structure. The cable-end displacement is due partly to the cable extension under load and the rotation of the tower, which is directly related to the extension of the back-stay AF.

Let the tower rotation be ϕ, and consider a tension in FD producing a unit value of x_1.

$$\text{Force in back-stay AF} = \cos\alpha_D/\cos\alpha_A \sin\alpha_D$$

$$\text{Extension of back-stay} = \frac{\cos\alpha_D}{\cos\alpha_A \sin\alpha_D}\left(\frac{l}{AE}\right)_{AF} = h\phi\cos\alpha_A$$

Movement of cable-end D vertically,

$$= l_{FD}\,\phi\cos\alpha_D + \left(\frac{l}{AE}\right)_{FD}\frac{1}{\sin^2\alpha_D}$$

$$= \frac{\cot^2\alpha_D}{\cos^2\alpha_A}\left(\frac{l}{AE}\right)_{AF} + \left(\frac{l}{AE}\right)_{FD}\frac{1}{\sin^2\alpha_D} = e_{11}$$

vertical movement of cable-end C due to tower rotation

$$\frac{\cot\alpha_C\cot\alpha_D}{\cos^2\alpha_A}\left(\frac{l}{AE}\right)_{AF}$$

$$= e_{21}$$

Similarly, for unit value of x_2

$$e_{22} = \frac{\cot^2\alpha_C}{\cos^2\alpha_A}\left(\frac{l}{AE}\right)_{AF} + \left(\frac{l}{AE}\right)_{FC}\frac{1}{\sin^2\alpha_C}$$

and

$$e_{12} = e_{21}$$

Thus, the compatibility condition at D is

$$f_{11}x_1 + f_{12}x_2 + f_{13}x_3 + u_1 = -e_{11}x_1 - e_{12}x_2$$

Similarly, at C

$$f_{21}x_1 + f_{22}x_2 + f_{23}x_3 + u_2 = -e_{21}x_1 - e_{22}x_2$$

and at B

$$f_{31}x_1 + f_{32}x_2 + f_{33}x_3 + u_3 = 0$$

These equations may be written

$$\left\{\begin{bmatrix} f_{11} & f_{12} & f_{13} \\ f_{21} & f_{22} & f_{23} \\ f_{31} & f_{32} & f_{33} \end{bmatrix} + \begin{bmatrix} e_{11} & e_{12} & 0 \\ e_{21} & e_{22} & 0 \\ 0 & 0 & 0 \end{bmatrix}\right\}\begin{bmatrix} x_1 \\ x_2 \\ x_3 \end{bmatrix} = -\begin{bmatrix} u_1 \\ u_2 \\ u_3 \end{bmatrix} \qquad (8.12)$$

The program described in section 8.2 can be used to perform this analysis simply by adding the cable flexibilities e to the girder flexibilities f before entering the matrix inversion procedure. Again it should be noted that, in the calculation of influence line ordinates, the matrices **X** and **U** in eq. (8.12) contain a number of columns equal to the number of unit load positions required.

8.4. Shear Wall Analysis by the Finite Element Method

The analysis of a shear wall, of the type commonly found in tall multi-storey buildings, can be accomplished very simply using the finite element method with digital computer. A convenient finite element in this case is the rectangular, plane stress, element developed in section 6.4. The elevation of the wall is divided into a suitable pattern of rectangular elements, depending on the storey-heights and layout of openings in the wall. A reasonably accurate analysis requires a fairly fine division of the wall, particularly in the neighbourhood of any openings, and it is essential to economize on computer storage as much as possible. The number of unknowns to be handled will frequently be of the order of several thousands, and an efficient band matrix solution procedure is absolutely essential.

Consider the wall shown in Fig. 8.5a, divided into a pattern of rectangular elements with m elements horizontally and n vertically. The nodes, and nodal displacements, are numbered across the width of the wall as shown in the figure. This ordering of the nodal displacements produces the minimum band width in the stiffness matrix, providing $n > (m + 1)$, which will normally be the case.

Any openings in the wall are represented by finite elements having zero thickness. This approach maintains a uniform band width enabling the standard band matrix solution programs of chapter 7 to be used, and permits any pattern of openings to be handled, whether these form a regular or irregular layout.

The rectangular, plane stress, finite element of section 6.4 is used, but first a re-arrangement of rows and columns of the stiffness matrix, eq. (6.29), is carried out so that the nodal displacements are in the order

$$\{u_i \quad v_i \quad u_j \quad v_j \quad u_k \quad v_k \quad u_l \quad v_l\}$$

The stiffness matrix of the pth element is then partitioned in the following way:

$$\mathbf{k}_p = \left[\begin{array}{c|c|c|c} (\mathbf{a}_1)_p & & \text{Symmetric} & \\ \hline (\mathbf{b}_1)_p & (\mathbf{a}_2)_p & & \\ \hline (\mathbf{c}_1)_p & (\mathbf{b}_2)_p & (\mathbf{a}_2)_p & \\ \hline (\mathbf{d}_1)_p & (\mathbf{c}_1{}^T)_p & (\mathbf{b}_1{}^T)_p & (\mathbf{a}_1)_p \end{array}\right] \qquad (8.13)$$

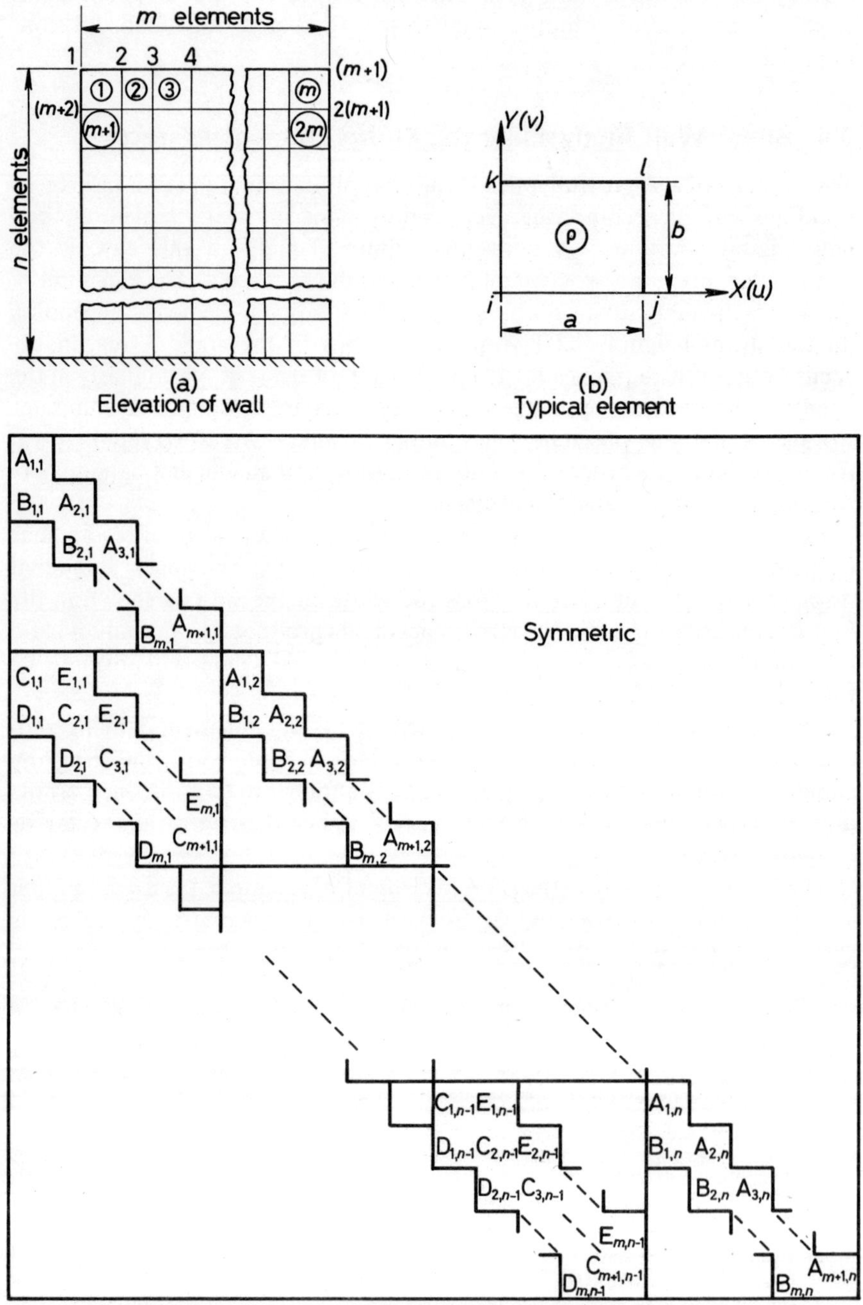

(c) Stiffness matrix (all unidentified submatrices are null)

Figure 8.5. Shear wall represented by finite element

where

$$\mathbf{a}_1 = \gamma \begin{bmatrix} 4(b^2 + \lambda_1 a^2) & 3ab\lambda_2 \\ 3ab\lambda_2 & 4(a^2 + \lambda_1 b^2) \end{bmatrix}$$

$$\mathbf{a}_2 = \gamma \begin{bmatrix} 4(b^2 + \lambda_1 a^2) & -3ab\lambda_2 \\ -3ab\lambda_2 & 4(a^2 + \lambda_1 b^2) \end{bmatrix}$$

$$\mathbf{b}_1 = \gamma \begin{bmatrix} 2(\lambda_1 a^2 - 2b^2) & 3ab(\lambda_1 - v) \\ 3ab(v - \lambda_1) & 2(a^2 - 2\lambda_1 b^2) \end{bmatrix} \qquad (8.14)$$

$$\mathbf{b}_2 = \gamma \begin{bmatrix} -2(b^2 + \lambda_1 a^2) & 3ab\lambda_2 \\ 3ab\lambda_2 & -2(a^2 + \lambda_1 b^2) \end{bmatrix}$$

$$\mathbf{c}_1 = \gamma \begin{bmatrix} 2(b^2 - 2\lambda_1 a^2) & 3ab(v - \lambda_1) \\ 3ab(\lambda_1 - v) & 2(\lambda_1 b^2 - 2a^2) \end{bmatrix}$$

$$\mathbf{d}_1 = \gamma \begin{bmatrix} -2(b^2 + \lambda_1 a^2) & -3ab\lambda_2 \\ -3ab\lambda_2 & -2(a^2 + \lambda_1 b^2) \end{bmatrix}$$

where

$$\gamma = \frac{Et}{12ab(1 - v^2)}; \quad \lambda_1 = \tfrac{1}{2}(1 - v); \quad \lambda_2 = \tfrac{1}{2}(1 + v)$$

The form of the stiffness matrix for the complete wall is easily defined once the pattern developing on summation of stiffnesses at the nodes is recognized. The stiffness matrix is as shown in Fig. 8.5c and all the sub-submatrices are 2×2 matrices. The sub-submatrices are defined in Table 8.3.

The program allows variations in horizontal and vertical dimensions of the rectangular elements, and in wall thickness, thus any pattern of openings can be accommodated by alloting zero thickness to the elements in the position of the opening. When all the elements meeting at a node have zero thickness, and consequently zero stiffness, a zero appears on the diagonal of the stiffness matrix. The band matrix procedures of chapter 7 break down on encountering this zero. The situation can be avoided by giving small values to the element thickness in the opening. A thickness of 10^{-10} of the

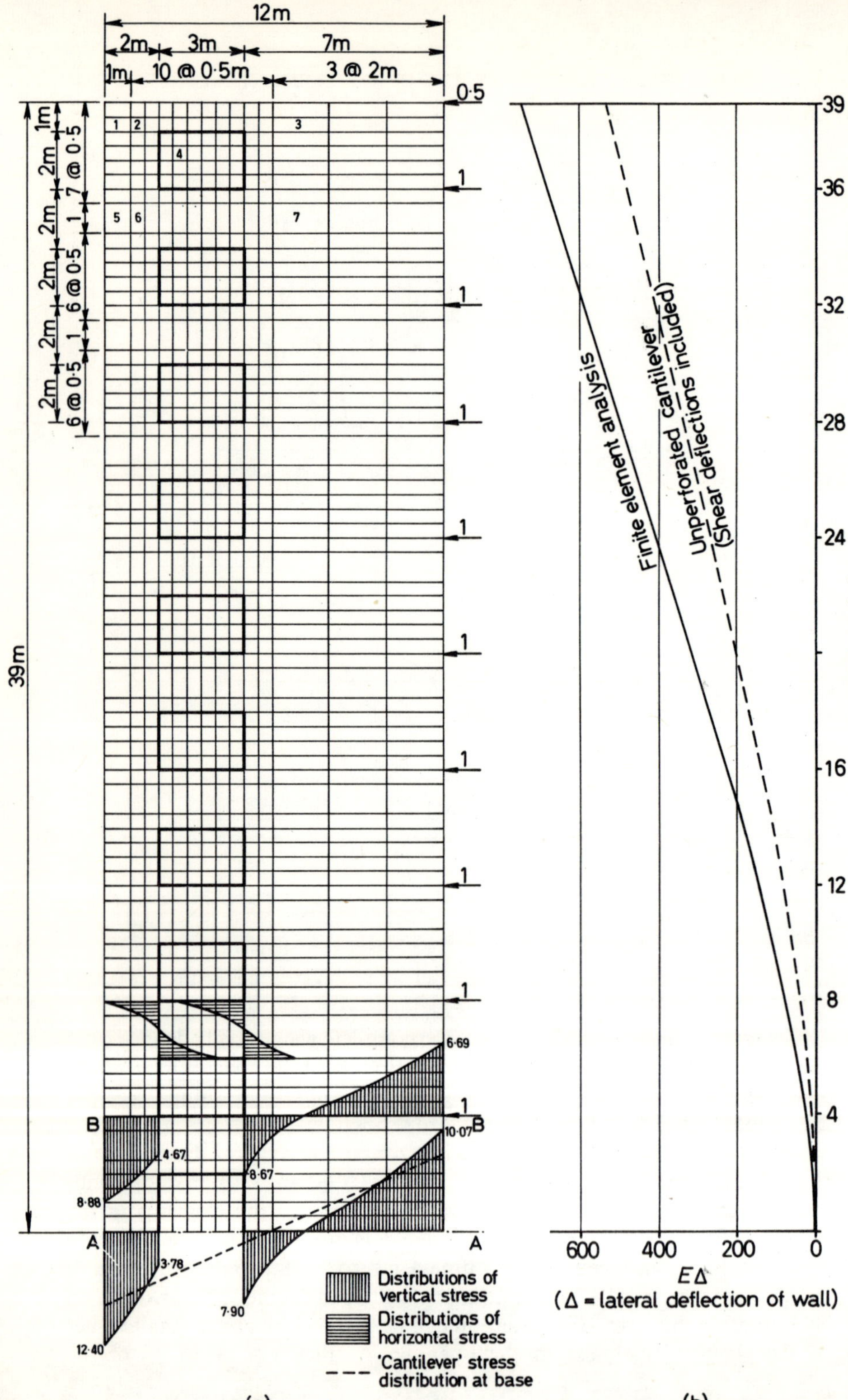

Figure 8.6. Finite element analysis of perforated shear wall $\left\{\begin{matrix} t = 1 \\ \nu = 0{\cdot}15 \end{matrix}\right\}$

neighbouring wall thickness is adequate for this purpose. An alternative procedure is to delete the row and column of the stiffness matrix corresponding to the zero diagonal element and repack the matrix.

Table 8.3. Assembly of Stiffness Matrix for Shear Wall Analysis

Sub-sub-matrices	s	r	Constituent Sub-matrices (all 2×2)
$\mathbf{A}_{r,s}$	1	1	$(a_2)_1$
		$1<r<(m+1)$	$(a_1)_{r-1}+(a_2)_r$
		$(m+1)$	$(a_1)_m$
	$1<s\leqslant n$	1	$(a_1)_{(s-2)m+1}+(a_2)_{(s-1)m+1}$
		$1<r<(m+1)$	$(a_1)_{(s-2)m+r}+(a_2)_{(s-2)m+r-1}$ $+(a_2)_{(s-1)m+r}+(a_1)_{(s-1)m+r-1}$
		$(m+1)$	$(a_2)_{(s-1)m}+(a_1)_{sm}$
$\mathbf{B}_{r,s}$	1	$1\leqslant r\leqslant m$	$(b_1{}^T)_r$
	$1<s\leqslant n$	$1\leqslant r\leqslant m$	$(b_1{}^T)_{(s-1)m+r}+(b_1)_{(s-2)m+r}$
$\mathbf{C}_{r,s}$	$1\leqslant s\leqslant(n-1)$	1	$(c_1{}^T)_{(s-1)m+1}$
		$1<r<(m+1)$	$(c_1)_{(s-1)m+r-1}+(c_1{}^T)_{(s-1)m+r}$
		$(m+1)$	$(c_1)_{(s-1)m}$
$\mathbf{D}_{r,s}$	$1\leqslant s\leqslant(n-1)$	$1\leqslant r\leqslant m$	$(b_2{}^T)_{(s-1)m+r}$
$\mathbf{E}_{r,s}$	$1\leqslant s\leqslant(n-1)$	$1\leqslant r\leqslant m$	$(d_1{}^T)_{(s-1)m+r}$

Some results obtained using the program are shown in Figs. 8.6a and b. The finite element results are compared with those obtained assuming the wall to be unperforated and to behave as a simple cantilever. The stress concentrations at the corners of the openings are clearly indicated by the finite element solution.

Bibliography

The following is a selected list of textbooks and published papers. For a more comprehensive bibliography the reader might consult the book by Przemieniecki.[31]

1. J. H. ARGYRIS and S. KELSEY, *Energy Theorems and Structural Analysis*, Butterworth (1960).
2. S. O. ASPLUND, *Structural Mechanics*, Prentice-Hall (1966).
3. S. BLASZKOWIAK and Z. KACZKOWSKI, *Iterative Methods in Structural Analysis*, Pergamon Press (1966).
4. D. M. BROTTON, *The Application of Digital Computers to Structural Engineering Problems*, Spon (1962).
5. D. M. BROTTON and D. F. BROOKS, 'Computer System for Analysis of Large Frameworks', *Proceedings of the American Society of Civil Engineers, Journal of the Structural Division*, December (1964).
6. J. CLARKSON, *The Elastic Analysis of Flat Grillages*, Cambridge University Press (1965).
7. R. W. CLOUGH, I. P. KING, and E. L. WILSON, 'Structural Analysis of Multistorey Buildings', *Proceedings of the American Society of Civil Engineers, Journal of the Structural Division*, June (1964).
8. R. W. CLOUGH, 'The Finite Element Method in Structural Mechanics', *Stress Analysis*, Chapter 7, Wiley (1965).
9. D. J. DAWE, 'Parallelogrammic Elements in the Solution of Rhombic Cantilever Plate Problems, *Journal of Strain Analysis*, **1** (No. 3) (1966).
10. S. J. FENVES, *Computer Methods in Civil Engineering*, Prentice-Hall (1967).
11. R. A. FRAZER, W. J. DUNCAN, and A. R. COLLAR, *Elementary Matrices*, Cambridge University Press (1946).
12. J. M. GERE and W. WEAVER, *Analysis of Framed Structures*, Van Nostrand (1965).
13. M. S. GRĘGORY, *Linear Framed Structures*, Longmans (1966).
14. A. S. HALL and R. W. WOODHEAD, *Frame Analysis*, Wiley (1961).
15. DEN HARTOG, *Advanced Strength of Materials*, Chapter 1, McGraw-Hill (1952).
16. J. C. DE C. HENDERSON and W. G. BICKLEY, 'Statical Indeterminacy of a Structure', *Aircraft Engineering* **27**, 400–402 (1955).
17. L. R. HERRMANN, 'Elastic Torsional Analysis of Irregular Shapes', *Proceedings of the American Society of Civil Engineers, Engineering Mechanics Division*, December (1965).
18. J. IRVING and N. MULLINEUX, *Mathematics in Physics and Engineering*, Appendix 5.2, Academic (1959).

19. W. M. Jenkins, 'Influence Line Computations for Structures with Members of Varying flexural Rigidity using the Electronic Digital Computer', *The Structural Engineer*, September (1961).
20. W. M. Jenkins, 'Multi-storey Frame Analysis by Digital Computer', *Acier-Stahl-Steel*, July/August (1963).
21. W. M. Jenkins and T. Harrison, 'The Analysis of Tall Buildings with Shear Walls under Bending and Torsion', *Symposium on Tall Buildings, Southampton University: April 1966*, Pergamon (1967).
22. W. M. Jenkins and L. C. Carpenter, 'The Analysis of Large Structural Systems by Digital Computer using the Stiffness Method', *Symposium on the Use of Computers in Structural Analysis and Design*, University of Newcastle: July (1966).
23. W. M. Jenkins and A. G. Bellamy, 'Design Method for Shear Walls with Openings', *Civil Engineering and Public Works Review*, March (1968).
24. H. I. Laursen, *Matrix Analysis of Structures*, McGraw-Hill (1966).
25. R. K. Livesley, *Matrix Methods of Structural Analysis*, Pergamon (1964).
26. H. C. Martin, *Introduction to Matrix Methods of Structural Analysis*, McGraw-Hill (1966).
27. S. J. McMinn, *Matrices for Structural Analysis*, Spon (1962).
28. R. J. Melosh, 'Basis of Derivation of Matrices for the Direct Stiffness Method', *A.I.A.A. Journal*, July (1963).
29. P. B. Morice, *Linear Structural Analysis*, Thames & Hudson (1959).
30. L. A. Pipes, *Matrix Methods for Engineering*, Prentice-Hall (1963).
31. J. S. Przemieniecki, *Theory of Matrix Structural Analysis*, McGraw-Hill (1968).
32. M. F. Rubinstein, *Matrix Computer Analysis of Structures*, Prentice-Hall (1966).
33. B. Stafford Smith, 'The Single-plane Cable-stayed Bridge: A Method of Analysis Suitable for Computer Use', *Proceedings of the Institution of Civil Engineers*, May (1967).
34. B. Stafford Smith, 'A Linear Method of Analysis for Double-plane Cable-stayed Bridges', *Proceedings of the Institution of Civil Engineers*, January (1968).
35. S. S. Tezcan, 'Simplified Formulation of Stiffness Matrices', *Proceedings of the American Society of Civil Engineers, Journal of the Structural Division*, December (1963) (discussion on paper by P. M. Wright, April 1963).
36. S. S. Tezcan, 'Computer Analysis of Plane and Space Structures', *Proceedings of the American Society of Civil Engineers, Journal of the Structural Division*, April (1966).
37. S. Timoshenko, *Strength of Materials* (Part II: Advanced theory and problems, Chapter 7) 3rd edition, Van Nostrand (1956).
38. S. Timoshenko and S. Woinowsky Krieger, *Theory of Plates and Shells*, McGraw-Hill (1959).
39. P. C. Wang, *Numerical and Matrix Methods in Structural Mechanics*, Wiley (1966).
40. O. C. Zienkiewicz and Y. K. Cheung, 'The Finite Element Method for Analysis of Elastic, Isotropic, and Orthotropic Slabs', *Proceedings of the Institution of Civil Engineers*, August (1964).
41. O. C. Zienkiewicz and G. S. Hollister, *Stress Analysis*, Wiley (1965).
42. O. C. Zienkiewicz, *The Finite Element Method in Structural and Continuum Mechanics*, McGraw-Hill (1967).

Additional references on numerical analysis and the solution of linear simultaneous equations:

43. 'Modern Computing Methods', *Notes on Applied Science No. 16*, 2nd edition, National Physical Laboratory, H.M.S.O. (1961).
44. I. S. BEREZIN and N. P. ZHIDKOV, *Computing Methods*, Vol. 2, Chapter 6, Pergamon (1965).
45. L. DALE HARRIS, *Numerical Methods using Fortran*, Merrill (1964).
46. D. D. MCCRACKEN and W. S. DORN, *Numerical Methods and Fortran Programming*, Wiley (1965).
47. A. RALSTON, *A First Course in Numerical Analysis*, Chapter 9, McGraw-Hill (1965).

Appendix A

m_i \ m_j	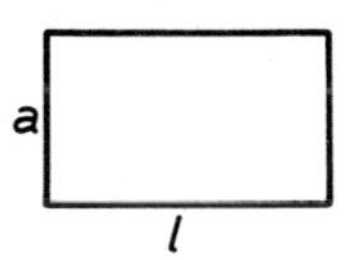	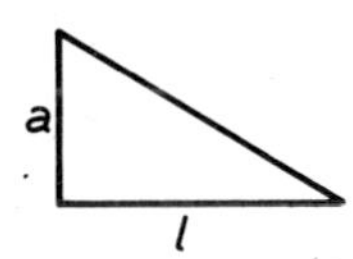	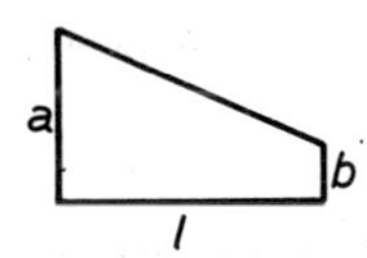
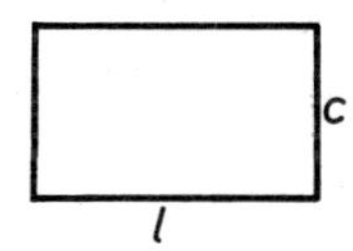	lac	$\frac{l}{2}ac$	$\frac{l}{2}(a+b)c$
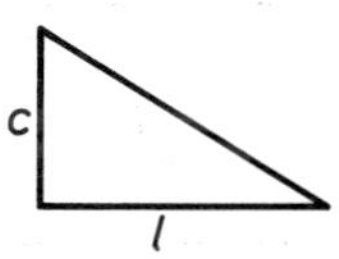	$\frac{l}{2}ac$	$\frac{l}{3}ac$	$\frac{l}{6}(2a+b)c$
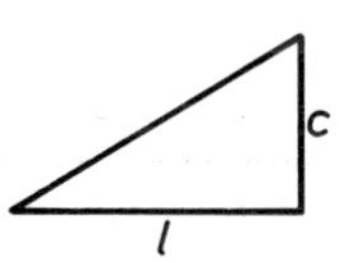	$\frac{l}{2}ac$	$\frac{l}{6}ac$	$\frac{l}{6}(a+2b)c$
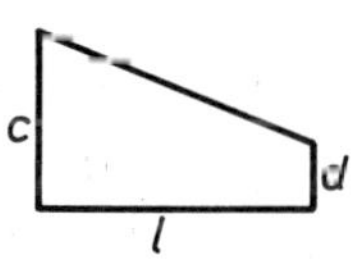	$\frac{l}{2}a(c+d)$	$\frac{l}{6}a(2c+d)$	$\frac{l}{6}\{a(2c+d)+b(2d+c)\}$
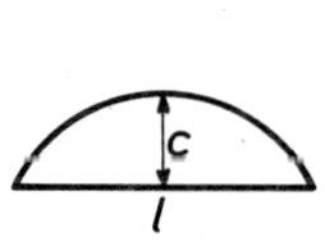	$\frac{2}{3}lac$	$\frac{l}{3}ac$	$\frac{l}{3}(a+b)c$

Appendix A. Table of product integrals for use in the flexibility method

$$\int_0^l m_i\, m_j\, ds$$

Appendix B

Loading	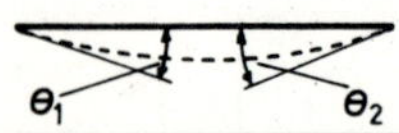		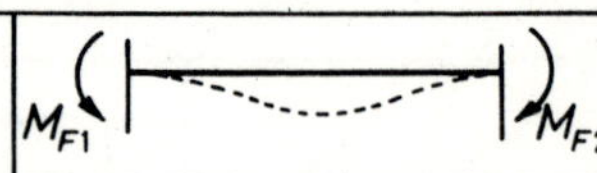	
	θ_1	θ_2	M_{F1}	M_{F2}
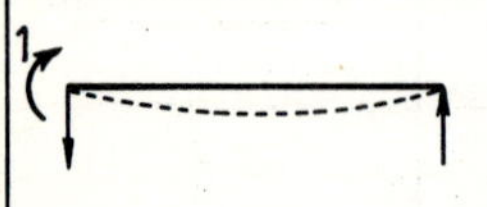	$\frac{l}{3EI}$	$\frac{l}{6EI}$		
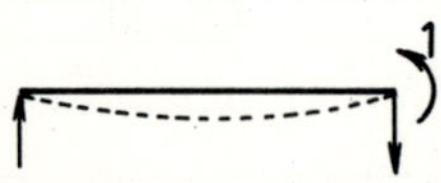	$\frac{l}{6EI}$	$\frac{l}{3EI}$		
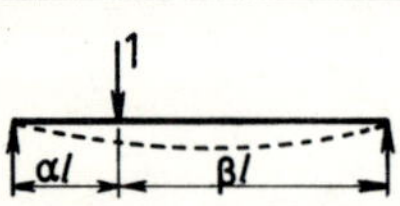	$\frac{l^2}{6EI}\beta(1-\beta^2)$	$\frac{l^2}{6EI}\alpha(1-\alpha^2)$	$l\alpha\beta^2$	$l\beta\alpha^2$
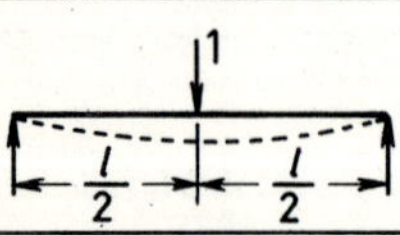	$\frac{l^2}{16EI}$	$\frac{l^2}{16EI}$	$\frac{l}{8}$	$\frac{l}{8}$
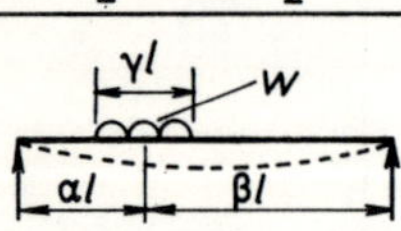	$\frac{w\beta\gamma l^3}{24EI}\left[4\alpha(1+\beta)-\gamma^2\right]$	$\frac{w\alpha\gamma l^3}{24EI}\left[4\beta(1+\alpha)-\gamma^2\right]$	$\frac{wl^2\gamma}{12}\left[12\alpha\beta^2-\gamma^2(2\beta-\alpha)\right]$	$\frac{wl^2\gamma}{12}\left[12\beta\alpha^2-\gamma^2(2\alpha-\beta)\right]$
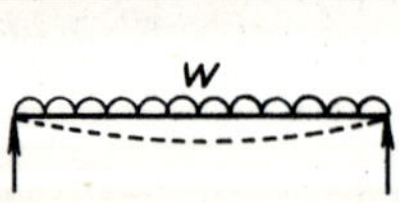	$\frac{wl^3}{24EI}$	$\frac{wl^3}{24EI}$	$\frac{wl^2}{12}$	$\frac{wl^2}{12}$
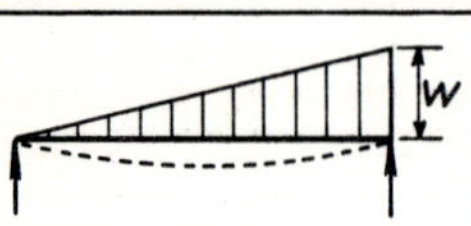	$\frac{7}{360}\ \frac{wl^3}{EI}$	$\frac{wl^3}{45EI}$	$\frac{wl^2}{30}$	$\frac{wl^2}{20}$

In all cases, $M_{F1} = \frac{2EI}{l}(2\theta_1 - \theta_2)$

$M_{F2} = \frac{2EI}{l}(2\theta_2 - \theta_1)$

Appendix B. End rotations and fixed-end moments for prismatic beams

Appendix C

Section	A	I_X	I_Y	J †
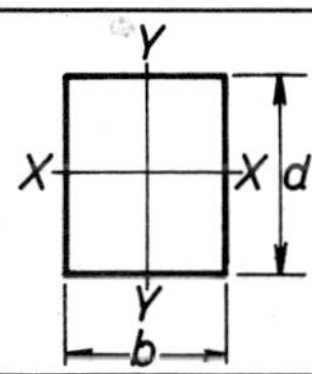	bd	$\frac{bd^3}{12}$	$\frac{db^3}{12}$	kdb^3 ‡ $(d \geqslant b)$
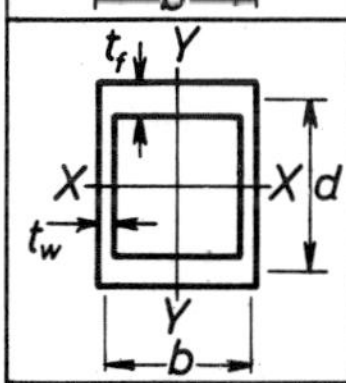	$2(bt_f+dt_w)$	$\frac{d^2}{6}(t_w d+3t_f b)$ *	$\frac{b^2}{6}(t_f b+3t_w d)$ *	$\frac{2b^2d^2t_w t_f}{bt_w+dt_f}$ *
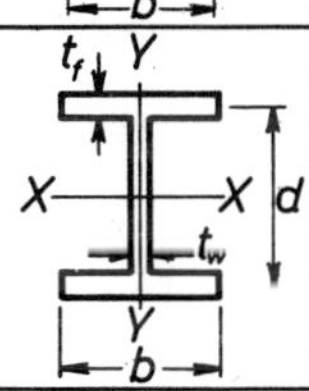	$2bt_f+dt_w$ *	$\frac{d^2}{12}(6bt_f+dt_w)$ *	$\frac{t_f b^3}{6}$ *	$\frac{1}{3}(dt_w^3+2bt_f^3)$ *
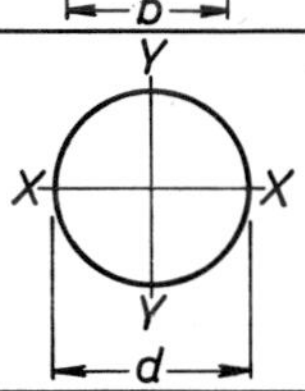	$\frac{\pi d^2}{4}$	$\frac{\pi d^4}{64}$	$\frac{\pi d^4}{64}$	$\frac{\pi d^4}{32}$
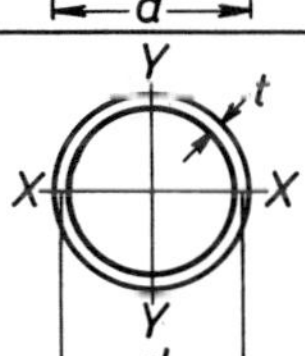	πdt *	$\frac{\pi d^3 t}{8}$ *	$\frac{\pi d^3 t}{8}$ *	$\frac{\pi d^3 t}{4}$ *

* Approximate for 'thin-walled' members

† Expressions are based on uniform torsion (no warping restraint)[37,15,17].

‡ $k \simeq \frac{1}{3} - 0{\cdot}21\,\frac{b}{d}\left(1 - \frac{b^4}{12d^4}\right)$ [REF. (12)]

Appendix C
Section properties

Answers to Problems

Chapter 1

1.1 $\mathbf{F} = \dfrac{l}{6EI}\begin{bmatrix} 4 & 1 \\ 1 & 4 \end{bmatrix}$

$$M_B = \frac{Wab}{15l^2}(2a + 7b);\quad M_C = \frac{Wab}{15l^2}(2b + 7a);\quad M_W = -\frac{Wab}{15l^3}(8l^2 + 10ab)$$

1.2 $M_{AB} = 157$ kNm; $M_{BA} = -121{\cdot}8$ kNm; $M_{CD} = 150{\cdot}9$ kNm

1.3 $\mathbf{K} = 24\dfrac{EI}{h^3}\begin{bmatrix} 1 & -1 \\ -1 & 2 \end{bmatrix};\quad \mathbf{F} = \dfrac{h^3}{24EI}\begin{bmatrix} 2 & 1 \\ 1 & 1 \end{bmatrix}$

1.4 $t_{BD} = \dfrac{Wa}{3\sqrt{2l}};\quad t_{BE} = \dfrac{Wa}{3\sqrt{2l}}$

$$M_{AB} = -\frac{Wa}{12l^3}(3ab + 2l^2 + 12b^2);\quad M_{BA} = \frac{Wa}{6l^2}(3ab - l^2);$$

$$M_{BC} = \frac{Wa}{6l^2}(l^2 - 3ab);\quad M_{CB} = \frac{Wa}{12l^2}(2l^2 - 3ab)$$

1.5 $\mathbf{K} = EI\begin{bmatrix} 4{\cdot}60 & 1{\cdot}65 \\ 1{\cdot}65 & 8{\cdot}54 \end{bmatrix}$

$$\theta_C = 0{\cdot}274\frac{w}{EI};\quad \theta_D = -0{\cdot}0925\frac{w}{EI}$$

$M_A = 1{\cdot}77w$ Nm; $M_B = 0{\cdot}595w$ Nm

1.6 $\mathbf{T} = \begin{bmatrix} \cos\alpha & \sin\alpha & 0 \\ -\sin\alpha & \cos\alpha & 0 \\ 0 & 0 & 1 \end{bmatrix}$

Chapter 2

2.1 $M_D = -60{\cdot}3$ kNm; $H_D = 31{\cdot}4$ kN; $V_D = 59{\cdot}6$ kN

2.2 $V_A = \frac{5}{3}$ kN; $V_C = \frac{55}{2}$ kN; $V_D = \frac{25}{2}$ kN; $V_B = -\frac{5}{3}$ kN;
$H_C = H_D = 15$ kN; $M_{FA} = -15$ kNm;
$M_{FE} = \frac{195}{4}$ kNm; $M_{GE} = -\frac{195}{4}$ kNm; $M_{GB} = 15$ kNm

2.3 $M_{DE} = \frac{6}{13}W$; $M_{ED} = -\frac{7}{13}W$

2.4 $M_{AC} = 5{\cdot}448$ kNm; $M_{CA} = 21{\cdot}804$ kNm; $M_{CD} = -21{\cdot}804$ kNm
$M_{DC} = 27{\cdot}276$ kNm; $M_{DB} = -27{\cdot}276$ kNm
$r_1 = 32{\cdot}7/EI$; $r_2 = 28{\cdot}5/EI$; $r_3 = 29{\cdot}1/EI$; $r_4 = 25{\cdot}4/EI$

2.5

$$\mathbf{F} = \frac{1}{12EI}\begin{array}{|c|c|c|c|c|c|c|c|c|}
\hline
8h+6l & & & & & & & & \\
\hline
-4h & 8h+6l & & & & & & & \\
\hline
-6h-3l & 0 & 12h+4l & & & & \text{Symmetric} & & \\
\hline
6h & -6h-3l & 0 & 12h+4l & & & & & \\
\hline
-4h-3l & 2h & 6h+4l & 0 & 16h+4l & & & & \\
\hline
8h+3l & -4h-3l & -6h-l & 6h+4l & -4h-l & 8h+8l & & & \\
\hline
-8h-3l & 4h+3l & 6h+4l & -6h-l & 12h+4l & -8h+2l & 16h+8l & & \\
\hline
4h+\frac{3}{2}l & -2h-\frac{3}{2}l & -3h-3l & 3h+l & -8h-3l & 4h+l & -10h-4l & 14h+3l & \\
\hline
4h+\frac{3}{2}l & -2h-\frac{3}{2}l & -3h-3l & 3h+l & -4h-3l & 4h+l & -6h-4l & 8h+3l & 10h+3l \\
\hline
\end{array}$$

2.6 $T_{BC} = 1{\cdot}14W$; $T_{BD} = 1{\cdot}37W$; $T_{BE} = 1{\cdot}24W$

Chapter 3

3.1 $\mathbf{F} = \dfrac{1}{EI_0}\begin{bmatrix} 0{\cdot}75 & 0{\cdot}50 \\ 0{\cdot}50 & 1{\cdot}25 \end{bmatrix}$

3.2 $\mathbf{F} = \dfrac{l}{EI_0}\begin{bmatrix} 0{\cdot}193 & 0{\cdot}0569 \\ 0{\cdot}0569 & 0{\cdot}0681 \end{bmatrix}$

$EI_0 = bd_0{}^3/12$

3.3 $f_{11} = f_{22} = \dfrac{23}{60}\dfrac{l}{EI}$

$f_{12} = f_{21} = \dfrac{7}{60}\dfrac{l}{EI}$

3.4 $M_A = -2.36W$; $M_B = 4{\cdot}50W$;
$H = 1{\cdot}50W$ (using non-linear interpolation for α in table 3.2)

3.5 $M_A = \frac{41}{92}WL$; $T_A = \frac{5}{92}WL$
$V_A = \frac{41}{92}W$; $V_D = \frac{51}{92}W$

3.6 $\frac{13}{23}WL$

3.7 $\mathbf{F} = 10^{-9}\begin{bmatrix} 0{\cdot}03638 & 0{\cdot}01674 & 0 \\ 0{\cdot}01674 & 0{\cdot}04821 & 0{\cdot}01674 \\ 0 & 0{\cdot}01674 & 0{\cdot}03638 \end{bmatrix}$

$$\mathbf{U} = -10^{-9}\begin{bmatrix} 0{\cdot}05178 & 0{\cdot}08246 & 0{\cdot}08214 & 0{\cdot}05242 & 0{\cdot}1286 & 0{\cdot}1990 & 0{\cdot}1805 & 0{\cdot}09908 \\ 0 & 0 & 0 & 0 & 0{\cdot}09908 & 0{\cdot}1805 & 0{\cdot}1990 & 0{\cdot}1286 \\ 0 & 0 & 0 & 0 & 0 & 0 & 0 & 0 \end{bmatrix}$$

$$\mathbf{X} = \begin{bmatrix} 1{\cdot}758 & 2{\cdot}799 & 2{\cdot}788 & 1{\cdot}779 & 2{\cdot}977 & 4{\cdot}224 & 3{\cdot}335 & 1{\cdot}559 \\ -0{\cdot}7264 & -1{\cdot}157 & -1{\cdot}152 & -0{\cdot}7354 & 1{\cdot}216 & 2{\cdot}710 & 3{\cdot}535 & 2{\cdot}531 \\ 0{\cdot}3343 & 0{\cdot}5324 & 0{\cdot}5303 & 0{\cdot}3384 & -0{\cdot}5595 & -1{\cdot}247 & -1{\cdot}627 & -1{\cdot}165 \end{bmatrix}$$

Chapter 4

4.1

$\mathbf{K} = EI$

$8/l + 4/h$								
$4/l$	$16/l + 4/h$							
0	$4/l$	$8/l + 4/h$				Symmetric		
$-6/h^2$	$-6/h^2$	$-6/h^2$	$36/h^3$					
$2/h$	0	0	$-6/h^2$	$8/l + 8/h$				
0	$2/h$	0	$-6/h^2$	$4/l$	$16/l + 4/h$			
0	0	$2/h$	$-6/h^2$	0	$4/l$	$8/l + 8/h$		
$6/h^2$	$6/h^2$	$6/h^2$	$-36/h^3$	0	$6/h^2$	0	$60/h^3$	
$-12/l^2$	0	$12/l^2$	0	$-12/l^2$	0	$12/l^2$	0	$96/l^3$

4.2

$\mathbf{K} =$

6470	1876	−470	0	0	0
1876	22000	−1876	0	0	0
−470	−1876	3470	0	0	0
0	0	0	6470	1876	1876
0	0	0	1876	22000	5000
0	0	0	1876	5000	16000

$$\mathbf{R} = \begin{bmatrix} 240 \\ 320 \\ 240 \\ 0 \\ 0 \\ 0 \end{bmatrix} \qquad \mathbf{r} = \begin{bmatrix} 0{\cdot}0379 \\ 0{\cdot}0185 \\ 0{\cdot}0846 \\ 0 \\ 0 \\ 0 \end{bmatrix}$$

$M_A = -111$; $M_B = 163$; $M_C = 111$ kNm

4.3

$$\mathbf{K} = \frac{EI}{100}$$

54·2					
12·5	300		Symmetric		
12·5	50	300			
37·5	50	0	200		
37·5	0	50	50	200	
−37·5	−37·5	−37·5	−37·5	−37·5	37·5

$$\mathbf{r} = \frac{1}{EI}\begin{bmatrix} 265{\cdot}5 \\ 28{\cdot}66 \\ 28{\cdot}66 \\ 9{\cdot}809 \\ 9{\cdot}809 \\ 369{\cdot}1 \end{bmatrix} \qquad \mathbf{S} = \begin{bmatrix} -52{\cdot}36 \\ -37{\cdot}70 \\ 15{\cdot}0 \\ -52{\cdot}36 \\ -37{\cdot}70 \\ 15{\cdot}0 \\ 42{\cdot}99 \\ 42{\cdot}99 \\ -10{\cdot}75 \\ -5{\cdot}298 \\ -14{\cdot}71 \\ 5{\cdot}00 \\ -5{\cdot}298 \\ -14{\cdot}71 \\ 5{\cdot}00 \\ 14{\cdot}71 \\ 14{\cdot}71 \\ -3{\cdot}68 \end{bmatrix} \begin{matrix} \left.\begin{matrix} \\ \\ \\ \end{matrix}\right\} 1, 3 \\ \left.\begin{matrix} \\ \\ \\ \end{matrix}\right\} 2, 4 \\ \left.\begin{matrix} \\ \\ \\ \end{matrix}\right\} 3, 4 \\ \left.\begin{matrix} \\ \\ \\ \end{matrix}\right\} 3, 5 \\ \left.\begin{matrix} \\ \\ \\ \end{matrix}\right\} 4, 6 \\ \left.\begin{matrix} \\ \\ \\ \end{matrix}\right\} 5, 6 \end{matrix}$$

4.4 $M_{AB} = 0{\cdot}183WL$; $M_{BA} = 0{\cdot}168WL$

4.5 0·0918 m; 5·8%

Chapter 5

5.1 $\mathbf{K} = EI_0\begin{bmatrix} 1{\cdot}82 & 0{\cdot}727 \\ 0{\cdot}727 & 1{\cdot}09 \end{bmatrix}$

5.2 $\mathbf{K} = \dfrac{EI_0}{l}\begin{bmatrix} 6{\cdot}86 & 5{\cdot}72 \\ 5{\cdot}72 & 19{\cdot}44 \end{bmatrix}$

5.3 $k_{AA} = k_{BB} = 0{\cdot}673EI$
$k_{AB} = k_{BA} = 0{\cdot}272EI$
$M_{FAB} = 0{\cdot}695W$
$M_{FBA} = 0{\cdot}506\text{W}$

5.4

$$\mathbf{K} = EI \left[\begin{array}{cccc|cc} \frac{4\beta_1}{a}+\frac{4}{c} & & & & \multicolumn{2}{c}{\text{Symmetric}} \\ -\frac{6\beta_1}{a^2}+\frac{6}{c^2} & \frac{12\beta_1}{a^3}+\frac{12}{c^3} & & & & \\ \frac{2}{c} & \frac{6}{c^2} & \frac{4}{c}+\frac{4\beta_2}{b} & & & \\ -\frac{6}{c^2} & -\frac{12}{c^3} & -\frac{6}{c^2}+\frac{6\beta_2}{b^2} & \frac{12}{c^3}+\frac{12\beta_2}{b^3} & & \\ \hline \frac{2\beta_1}{a} & -\frac{6\beta_1}{a^2} & 0 & 0 & \frac{4\beta_1}{a} & \\ 0 & 0 & \frac{2\beta_2}{b} & \frac{6\beta_2}{b^2} & 0 & \frac{4\beta_2}{b} \end{array}\right]$$

$$= EI \left[\begin{array}{c|c} \mathbf{K}_1 & \mathbf{K}_3{}^T \\ \hline \mathbf{K}_3 & \mathbf{K}_2 \end{array}\right]$$

$$\overline{\mathbf{K}} = \mathbf{K}_2 - \mathbf{K}_3\mathbf{K}_1{}^{-1}\mathbf{K}_3{}^T$$

5.5 $\overline{\mathbf{K}} = \dfrac{EI}{l}\begin{bmatrix} 14{\cdot}1 & -18{\cdot}3 \\ -18{\cdot}3 & 41{\cdot}3 \end{bmatrix}$

5.6 Displacements

Node	$10^6\ \theta_X$	$10^6\ \theta_Y$	$10^3 w$
1	12·62	16·93	0·000
2	22·19	−8·145	3·245
3	20·44	−20·44	0·000
4	4·997	20·37	1·951
5	7·665	−7·665	6·803
6	8·145	−22·19	3·245
7	−10·78	10·78	0·000
8	−20·37	−4·997	1·951
9	−16·93	−12·62	0·000

Member forces

Beam A–B	M_{XA}	M_{YA}	V_A	M_{XB}	M_{YB}	V_B
1–2	−3·736	2·681	−0·1079	3·736	−45·86	0·1079
2–3	1·368	46·23	0·2243	−1·368	−1·368	−0·2243
4–5	−1·041	−6·420	−0·1677	1·041	−60·68	0·1677
5–6	−0·3741	61·35	0·3323	0·3741	5·104	−0·3323
7–8	3·740	3·740	−0·0576	−3·740	−26·80	0·0576
8–9	−2·681	25·76	0·1101	2·681	−3·736	−0·1101
4–1	2·681	25·76	0·1101	−2·681	−3·736	−0·1101
5–2	0·3741	61·35	0·3323	−0·3741	5·104	−0·3323
6–3	−1·368	46·23	0·2243	1·368	−1·368	−0·2243
7–4	−3·740	3·740	−0·0576	3·740	−26·80	0·0576
8 5	1·041	−6·420	−0·1677	−1·041	−60·68	0·1677
9–6	3·736	2·681	−0·1079	−3·740	−45·86	0·1078

Index

THIS BOOK HAS BEEN SET IN MONOPHOTO TIMES NEW ROMAN
AND PRINTED AND BOUND IN GREAT BRITAIN BY
WILLIAM CLOWES AND SONS, LIMITED, LONDON AND BECCLES